FUNCTIONS AND OBJECTS
OF AUTHOR AND TITLE CATALOGUING

FUNCTIONS AND OBJECTS OF AUTHOR AND TITLE CATALOGUING

A CONTRIBUTION TO CATALOGUING THEORY

BY

Á. DOMANOVSZKY

ENGLISH TEXT EDITED BY

ANTHONY THOMPSON

VERLAG DOKUMENTATION · MÜNCHEN 1975

Eine Gemeinschaftsausgabe von Akadémiai Kiadó, Budapest
mit Verlag Dokumentation, München

Neugebunden 1975

ISBN 3-7940-4113-5

Printed in Hungary

CONTENTS

FOREWORD

More than anybody else, librarians are likely to be aware of the fact that – just as the worst of prefaces cannot for long prevent the reading public from taking notice of a good book – even the most masterly of them lacks the power to turn a bad book into a lasting success. By the very nature of their work they are supremely aware of the serious obstacles by which the information explosion of our times hampers the access to even the best books and therefore, it is more than ever their duty to remove any obstacles to that access by all the means at their disposal.

The most efficient, 'classical' tool which librarians have always been able to use in their pursuit of their aim, has been the catalogue. Even now the catalogue remains the most reliable intermediary between author and reader, it is still the most dependable guide by which the latter is led on his difficult route, a route beset and rendered in many respects uncertain by a large amount of historical, social and psychological contingencies.

If a chance still exists of mastering the unrestrained flood of information, it is quite certainly bound up with the help the librarians can contribute. Consequently, the importance of their catalogues and of the level of these catalogues being brought up to an optimum, is bound to steadily increase. There would obviously be neither rhyme nor reason in the automation of catalogues unless the data input reached the utmost reliability attainable; the machines would only exacerbate the harm done by the cataloguers' original mistakes.

In the following study Mr. Domanovszky submits with unrivalledly meticulous logic the central problems of the most fundamental section of all library work, that of author and title cataloguing, to a thorough examination. What he offers is no idle speculation on conception and nature of this branch of cataloguing, but – in the practical as well as the theoretical sense of the word – a penetrating analysis, based on a solid empirical foundation. It was not by mere chance that the author's name in library literature is internationally known. That in the Conferences of Paris 1961 and Copenhagen 1969 he was given an opportunity to give voice to his opinion and submit his proposals to assemblies of the world's most outstanding experts in author-title cataloguing, was the only natural outcome of his rich experience and his aptitude for acute critical analysis.

In this work Mr. Domanovszky, former Librarian of the University of Pécs, and Deputy Librarian of the oldest Hungarian research library, the Budapest University Library founded in 1635, presents an evaluation and generalization of a part of the experiences acquired in the course of a long-continued practical activity. Owing to the broad empirical basis of all that he has to say, there is not one among his propositions, even among those bearing the most explicitly theoretical character, which cannot be traced back to everyday cataloguing practice. The two principal merits of his inquiry consist in the strict logic of his theoretical analysis, and in the width of the experiential foundation upon which this analysis is based. In other words: the author is in possession of both of those vital qualities which alone can render the librarian a partner of equal rank to the research worker.

László Mátrai
President, Hungarian Library Association
Professor of Philosophy, University, Budapest
Librarian of the University, Budapest

PREFACE

The two premises from which I started when making the following inquiry were, first, that author-title cataloguing, being a very intricate technique, cannot do without an elaborate theoretical basis, and second, that what cataloguers have achieved by now in the way of building up such a theory is in many respects still rather far from satisfactory.

Accordingly, I have pursued two distinct ends. I wished first of all to make a positive attempt at improving the present state of theory, by making a thorough scrutiny of a particular section of it. The section I chose for this purpose is concerned with the functions and objects of author-title cataloguing. My reason for this choice has been that the functions of the catalogue constitute undoubtedly a really central, fundamental, and fascinating theme of cataloguing theory. It is only by analyzing its functions, its services, that one can disclose the essence of the catalogue; besides, its structure and the entire process and all the methods and devices of preparing it are – or certainly ought to be – determined by its functions. The ground for my extending the following investigations to the objects (not objectives!) of cataloguing too, is the fact that it is impossible to represent the functions adequately without inspecting minutely enough the objects about which the catalogue is communicating or is expected to communicate information. To deal with this twofold theme is, then, the first aim and the subject-matter proper of the following study.

Its second aim is to verify my two premises, that is, to prove that the present state of cataloguing theory *is* backward, and that this backwardness does entail detrimental consequences in practice. This aim I have tried to attain by pointing out – together with their practical consequences – every ill-chosen purpose, mistaken conception, erroneous notion, unclear concept, vague term, unsatisfactory rule, more or less current even in our own days, which I have come across in the course of discussing my subject-matter proper. It will be seen that, although the area examined is rather limited, the quantity of mistakes, that is, of the evidence for the validity of my second premise, is quite considerable.

Everything that concerns this second aim of mine will take, then, the form of illustrations of the subject-matter proper. This might make the reader inclined to think that all this is of tertiary importance only, and that the verification of those premises is an end entirely subordinate to my first aim and to my subject-matter proper. I wish to stress, therefore, that, in spite of their apparent subordination, the passages in question do form an integral part of what I have to say. So much so, that I should not assess my efforts as substantially abortive even if it proved true that I had succeeded only poorly in attaining my first aim provided that at the same time it also proved true that I contrived to demonstrate convincingly the validity of my two premises.

I cannot bring this introduction to a close before gratefully stating that this study would never have appeared without the efficient support of Professor Dr. László Mátrai, Chairman of the Section for Philosophy and History of the Hungarian Academy of Sciences and Librarian in Chief of the Eötvös Lóránd University of Budapest. His use of his influence was actually the sole motive force that was leading up to the publication of this book. I owe him a great debt for the time and trouble he so profusely and obligingly spent on obviating the obstacles, and on procuring the ways and means of publication.

I must also express my sincere thanks to Mr. Anthony Thompson, author of *Vocabularium Bibliothecarii* and former General Secretary of the International Federation of Library Associations, and to Miss Anikó Maleczky, editor of this book, for their valuable help given me by revising thoroughly and most competently my English text.

CHAPTER I

THE AIMS AND THE SUBJECT-MATTER OF THIS STUDY

1. IS THE CATALOGUE-WORKER AT LIBERTY, IF HE THINKS FIT, TO DISREGARD THE CATALOGUING CODE? "The recognition of cataloging as a truly intellectual discipline might be based" – according to L. Carnovsky – "apart from the construction of cataloging codes, on the fact that the cataloger has to exercise his judgement as to when codes and rules are to be disregarded"[1]. This view on the respective roles of code and cataloguer, or at least some of its implications, are shared by an appreciable portion of librarians, and this makes it worth while to take a closer look at it.

What may have induced Carnovsky to form this opinion? After all, consistency is one of the principal virtues a catalogue must possess, and consistency is, evidently, seriously endangered if the application or disregard of established rules becomes subject to the individual cataloguer's discretion. What can be gained by granting him this power which might outweigh the drawbacks it is obviously liable to entail? Does not Jewett's advice that "Nothing, as far as can be avoided, should be left to the individual taste or judgement of the cataloger"[2] definitively dispose of Carnovsky's view?

I can think of two explanations of his having taken it. The first is that he is an adherent of Cutter – a rather extreme one in holding that the catalogue must make allowances for the public's deeply-rooted habits of looking at things. After all, Cutter, in his capacity as a codifier, contemplated obviously only an adaptation of the rules of cataloguing to the public's expectations, by means of a set of explicitly stated exceptions from his general rules, but it is quite improbable that he should have thought of making his rules optional. So Carnovsky went considerably farther than Cutter did.

[1] CARNOVSKY, L.: Role of the Public Library: Implications for Library Education. *Library Quarterly*. Oct. 1964, p. 317. In spite of its irrelevance to my subject, I cannot refrain from voicing an objection also against another notion of Carnovsky revealed by this very same sentence: against his disparagement of normal cataloguing as an intellectual discipline. The reason for my not being able to abstain from doing so is that this is a belief which seems to be even more widespread among librarians than the one discussed in the text above. It is shared, obviously, by all those who have no direct personal knowledge of cataloguing work, as they have never been practising cataloguers themselves, or have parted with the job before having reached the stage of a tolerably good cataloguer.

[2] JEWETT, Ch. C.: *Smithsonian Report on the Construction of Catalogues of Libraries*. 2nd ed. Washington, Smithsonian Institution, 1853, p. 8.

I share the views of many contemporary authors on the Cutterian conception, and am convinced that for the public's own good one ought to be very cautious in establishing in codes exceptions under general rules with which users have already grown familiar. I consider a judicious employment of added entries a much more expedient method of accommodating the catalogue to specific expectations of the public which are contrary to an established general rule. Obviously, it is in an even more marked degree, that I must find this expedient preferable to any disregarding of codes and rules.

Apart from the wish to make concessions to the public's expectations, one may, however, also have another reason for maintaining Carnovsky's view on the cataloguer's rights and duties: the conviction that to give professional cataloguers a free hand to exercise their sound judgement is a corrective rendered directly indispensable by the inadequacy of the codes they have to apply. I cannot share this view either, seeing that the judgements of individual cataloguers, even if sound, are liable to vary greatly. If codes are inadequate, the remedy is to be sought, obviously, in their revision, and not in giving cataloguers a *carte blanche* to disregard them.

But are they really inadequate?

2. Codes are in fact far from being faultless. At any rate, it does not seem easy to find an experienced and at the same time thoughtful cataloguer who is content with all the rules of the code for author-title cataloguing on the basis of which he is obliged to work. This applies without exception to every code. Most of the complaints raised by cataloguers against codes have their good grounds. All the codes in use today have shortcomings of all sorts, beginning with simple superficialities originating in a negligence of the code-makers, as well as an astonishing lack of precision, clarity and articulateness in expressing their own ends and intentions; and ending with downright mistakes committed in the very choice of these aims.[3]

As codes form the basis of author-title cataloguing in hundreds or even thousands of libraries, so that every single mistake they commit entails a proliferation of mistakes in library catalogues, one would think, that conscious of this multiplying effect, code-makers take the utmost care that no mistakes should slip into their codes. Even a cursory skimming of the most modern codes suffices, however, to prove the contrary. We shall illustrate this by concrete examples later on. Meanwhile, let us point out a general evidence of the validity of this statement.

In our own days the different national codes are still very far from being uniform. Even those of countries with closely related civilization very often keep applying widely differing solutions to the very same cataloguing problems. Under present-day conditions of international cultural relations such a state of affairs is truly anachronistic. Author-title catalogues are tools; author-title cataloguing is an essentially technical art (though still a rather unmechanized one). But technologies can be based on rather solid foundations, and this implies that among several diverse solutions applied actually to a particular task, there must always be

[3] As to the *A. L. A. Rules* this has been brilliantly demonstrated by Seymour Lubetzky in his *Cataloging Rules and Principles. A Critique of the A. L. A. Rules for Entry and a Proposed Design for their Revision.* Washington, Library of Congress, 1953.

one which in the given state of technical development must be considered optimal by all judicious experts. Obviously, this must also apply to cataloguing: there must exist optimal solutions which ought to be adopted and enforced by all cataloguing codes, without regard to national boundaries. However, in spite of the Paris Conference of 1961, the progress towards international conformity is still astonishingly slow. The latest codes and drafts of codes persist in differing from one another even in cardinal matters. This means that most of them are committing the rather perplexing mistake of insisting on the application of suboptimal solutions; and this in spite of the fact that their compilers must, necessarily, have been aware of methods already developed and employed elsewhere the superiority of which they ought to recognize.

Such a presentation of the present-day cataloguing codification is seemingly open to the objection that the compilers of codes are obliged to take into account the cataloguing tradition and the conditions prevailing in their countries. Consequently, it is supposed, they are prevented by considerations of continuity from adopting alternatives the superiority of which they have already duly recognized. This objection is not valid. First, new codes are made just in order to introduce innovations. However, to make innovations, if one has already resolved to make them, only half-heartedly and piecemeal, is in cataloguing an utterly wrong policy. Secondly, consideration for venerable old practices and catalogues must not be permitted to prevent the introduction of newly-invented better methods into newly-established catalogues. It does not often occur either that compilers of codes account for their irrational decisions by pleading consideration for continuity and economic imperatives; it is a common experience that for each of several widely diverging solutions the character of optimality is claimed by their respective supporters.

Cataloguing is, thus, a technical activity which obviously has not yet reached the stage of development in which the majority of real experts easily agree, at least in fundamental matters, upon the solutions which have to be considered optimal, i. e., a stage that nowadays may already be considered normal in most other technological disciplines. Notions on cataloguing are, then, at present less advanced and more confused than the ideas which govern activities in the field of other technologies.

3. Which is more apt soon to bring about an improvement of codification – theory or statistics? How is this backward state of our art to be explained? And how can it be helped?

In our days many cataloguers maintain that the comparative appropriateness of the diverse methods actually employed in solving a particular task must be ascertained by statistical investigations. But these, if not carried out on a very large scale, do not produce conclusive results. Since under present conditions they are excessively cumbersome, and for psychological reasons also rather unreliable, there does not seem to be much hope of their being carried out before long on an adequate scale. True, by the advent of computerized cataloguing the relevant conditions would be altered radically, perhaps in a not too distant future. But this is a contingency which, in my opinion, does not warrant postponing all the endeavours to improve the present situation until the prospect of abundant computer-produced statistics on catalogue use became a reality. The less so, as the success of any statistical investigation depends on the indispensable condition of its having a reliable theoretical foundation. We are, however, in many respects incapable of providing such a foundation. That is why at the present moment even automated statistics would not be able to deliver us from all our evils.

In my opinion it is, then, not a lack of statistics, but the lack of a systematical theory of cataloguing, which is at the core of our present difficulties. We must strive, first of all, to remedy that; the more so, as this task we can attack without delay. The lack of an elaborate theory familiar to all cataloguers is responsible not only for the backward state of our cataloguing techniques, but also for the surprisingly uneven level of the theoretical papers published even in these days on questions of author-title cataloguing. Obviously, we have to take more trouble than heretofore to provide our craft with a theoretical foundation. That is to say, we should at last closely examine, first, our concepts and our terminology, and secondly, the relevant facts we have to deal with, including their interrelations and the possibilities of their classification.

The view that the terminology of author-title cataloguing is so imperfect as to be still greatly in need of improvement and completion, might appear at first glance a downright exaggeration. One might think that cataloguers having had several centuries for elaborating their terminology, it is quite improbable that this should not have yet reached a satisfactory stage of development.

As a matter of fact, cataloguers have *not* succeeded yet in creating a really adequate terminology. It is not so much new technical terms conveying to the outsider nothing that we are first of all in need of, although we cannot manage without introducing new terms of this sort either. Our chief terminological task to be solved is, however, to examine thoroughly the words of everyday language which we have lightheartedly adopted without adapting them to our purposes. Since words usually have in everyday language several meanings, the unequivocal denotation they are to bear in our professional language must always be fixed very carefully. This is a requirement not only plausible, but also self-evident, which remains, nevertheless, mostly unfulfilled. In our reasoning about problems of cataloguing we constantly keep stumbling over equivocal terms. For the validity of the thesis, maintained by semanticists, that words play a crucial part in moulding our thoughts, and for the deplorable consequences which an omission to act upon it may involve, our codes and the literature on cataloguing furnish many striking illustrations. In the following we, too, shall be able to point out cases in which errors in cataloguers' thinking, and the resulting defects in their rules and catalogues, have been brought about by the indiscriminate adoption of ambiguous words of common speech.

The second task mentioned above is, in just the same measure as the first, seemingly self-evident and therefore not worth mentioning. However, there are, rather surprisingly, not a few facts relevant to cataloguing which have never been examined thoroughly or stated explicitly either by theorists or by code-makers. One of the most important groups of these facts consists of the intentions and purposes of cataloguers, sometimes even of those concerning highly significant matters.

Present conditions in cataloguing practice prove that both these tasks must be solved urgently. Although the catalogue is only a tool, cataloguing merely a practical activity, a technical procedure, they cannot do without a theory. We must not forget that the tool as well as the procedure of its production is of a very complex nature. Nowadays in engineering it would not occur to anyone to set out to elaborate instructions for the large-scale manufacturing of a machine of similar intricacy without having an adequate theoretical basis to rely on. In this respect, too, cataloguers fall behind engineers considerably. The cause is probably that from the absolutely indisputable fact that cataloguing is a practical activity, they draw the completely false inference that it is not in need of any elaborate theory, practice being capable of finding the most appropriate ways spontaneously. This overrating of the role and competence of the catalogue-worker does not only logically imply, but has also practically led to, the neglect of cataloguing theory to which we refer, and also to an underestimate of the role and importance, and consequently to the factual inadequacy, of our codes.

4. The subject-matter and the two objectives of this study. The present study has two distinct objectives. It is, first of all, an attempt at outlining one of the chapters of an adequate cataloguing theory, a tentative first step towards elaborating that theory which, in our opinion, author-title cataloguing needs, but is not yet provided with. But at the same time it is also an attempt at proving the validity of the two propositions comprised in this opinion. This second end I shall try to attain by pointing out such examples of a not sound enough thinking current even nowadays which I have found in the course of following the first objective: unclear concepts and terms with vague meanings which cataloguers pertinaciously do stick to, as well as instances of a neglect thoroughly to examine and to clarify their own ends and the fitness of the means used for attaining them, or the objective facts which confront them when pursuing these ends, etc. Once or twice I shall even transgress the strict boundaries of my subject by a trifle, in order to be able to present especially conclusive evidence of the validity of those two propositions.

For that particular chapter of cataloguing theory the outlining of which is to be the first objective of this study, I have chosen one of the basic chapters of the theory of *author-title,* or *descriptive cataloguing* in the broader sense of the word[4], (the latter one is perhaps a more appropriate name which, however, seems gradually to be displaced by the former one) – the chapter on the functions and objects of this branch of cataloguing. This constitutes, then, the subject-matter proper of this study. Everything we shall say about it applies not only to the separate, detached author-title catalogue, but also to the author-title component of the dictionary, and even of an automated catalogue.[5] It should be borne in mind, therefore, that in what follows the term 'author-title catalogue' always includes also the author-title component of the dictionary and the author-title aspect of the automated catalogue.

In the choice of our subject-matter we have been led by the consideration that the structure of the author-title catalogue, the entire process and all the ways and means of author-title cataloguing depend on the functions to be performed by this

[4] When used in its narrower sense, the latter term denotes the preparation of the body, the imprint, the collation, and the notes of catalogue entries only, but excludes the choice of the heading and of its form. Lately the term has come to be used principally with this restricted denotation, as a rule without any application of qualificatory appendage. This accounts for the suppression of its use in the first, more comprehensive sense.

[5] Almost everything we shall have to say applies also to bibliographies arranged according to titles and author's names, if only for the words 'available in the library' the phrase 'recorded in the bibliography' is substituted.

catalogue. The scrutiny of these functions constitutes, thus, one of the most cardinal subjects of the theory of this branch of cataloguing. In the last few decades it was dealt with often enough, and there has also been found an answer to the question of functions which is rather widely accepted. This answer is, however, satisfactory in its main outlines only. It contains a lot of obscure points, as the matter has never been investigated in all its details. First of all, the scrutiny of the *functions*, the *objectives* of author-title cataloguing, has never been supplemented by a suitably detailed inspection of its *objects*.

By the term 'objects of cataloguing' I mean those objects about which information is conveyed by library catalogues, and *not the objectives of catalogues*, for which I propose consequently to use the words 'functions' or 'tasks'. Since we are concerned with the author and title catalogue only, in what follows we shall use the term 'objects of cataloguing', without mentioning this each time explicitly, to denote only those objects about which it is the task of the author-title catalogue to give information.

The topic of the functions is closely connected with that of the objects. You cannot even perfunctorily deal with the former without having at least to touch upon the latter, but if you want to give an accurate representation of the functions of the present-day author-title catalogue, you are also obliged to go into rather minute details concerning its objects. All the usual statements about the functions of the catalogue are concerned merely with their essential nature. They do not go beyond outlining crudely the contours of these functions, do not aim at an exhaustive and as far as possible correct characterization of them. This permitted the authors of these statements to confine themselves, so far as the objects of cataloguing are concerned, to highly generalizing propositions. Such a perfunctory handling of the matter cannot yield, however, a realistic representation, embracing all really significant particulars of the functions, the services which in these days are either actually performed, or for good reasons expected to be performed by the catalogue; only a sketchy, approximative, highly generalizing one. The traditional statements concerning the functions present these as if their boundaries were determined solely by their own intrinsic logic, and displayed regular, even shapes. The bonds of the area actually covered by author-title cataloguing do not correspond, however, to these *logical boundaries* of its functions. Their *real boundaries* are determined, along with economical considerations, by the actual needs of the catalogue-users, or more precisely, by the more or less sound notions of cataloguers on these needs. Therefore the regular shape, these boundaries ought to take logically, becomes often distorted. The scope of every function, as determined by its own intrinsic logic, contains sections where it is not worth while to fulfil it in practice. In such spots the boundaries of this scope, as established by practice, display deviations inwards from the boundaries indicated by logic. On the contrary, they bulge outwards whenever such a transgression of the logical boundaries seems to be justified by real demands. If we do not wish to be content with defining the functions approximatively, but want to record also the capricious curvatures of their boundaries, mostly generated spontaneously by cataloguing practice, we must supplement the scrutiny of the essence of the three functions by an examination and inventory of the objects they actually include. It is, however, impossible to give a complete account of all the objects, and in consequence, of the precise extent of the services of author-title cataloguing. We must be satisfied with an approximately, reasonably accurate inventory, which must not fail, however, to embrace at least all the significant types of objects. This should, naturally, not disappoint us, since the task of theory does not consist in giving a full account of reality. Yet, we must not leave off trying to make our scrutiny of the objects as thorough as possible, as our representation of the functions of the catalogue becomes the more realistic the further we advance in giving a detailed account of the objects they comprise.

But is it really by a specification of the objects that we can make our representation of the functions in the highest practicable degree accurate and realistic? Is this really the most suitable and practical way of attaining that end? Would it not be simpler and more obvious to attain it by an analysis of the functions themselves, by bringing to light the subordinate tasks the latter are composed of, that is to say, the tasks of cataloguing which belong to lower levels than the functions?

The answer is in the negative, simply because an inquiry into all these part-tasks would make a more comprehensive and more complex subject than the one is we propose to deal with. The latter, the scrutiny of the functions of cataloguing, necessarily includes, as already mentioned, a survey of all the significant types of objects they comprise. We may now add that it includes also the examination of the reasons for accepting the obligation of providing information about these types of objects. But it does not comprise the need to examine the details of the ways, the technical devices, which are most serviceable for recording these objects. If, however, we proposed to look into all the subordinate tasks the functions comprise, we should be obliged to examine – along with their objects and with the reasons for recording these objects – also all those methods and devices, that is to say, the questions of the choice and the form of the heading of the main entry as well as of the necessary added entries and references, in regard to each particular type of object separately, and even some questions of descriptive cataloguing in the narrower sense of the word. So we should not be spared the trouble of having to scrutinize, just as laboriously, the very same topic of objects, and should have to bother additionally also about the latter topics. The course we have chosen enables us, on the other hand, to confine ourselves to discussing only those technical means and devices in question which are entirely indispensable to discharging the functions. All the others, the services of which are confined to a mere raising of the efficiency with which the functions are getting discharged, we shall consider only if there is a peculiar reason for doing so. This applies first of all to the most controversial question of this area: the choice of the heading of the main entry when several entries are to be made for one object.[6]

The topic of the objects of cataloguing is much more neglected than that of the functions. So far as I can remember, I have never met with, either in the literature about cataloguing or in codes, any attempt at a systematical survey of these objects. This will render the following study to some extent disproportioned: we shall be obliged to concede to the topic of objects, in comparison with that of the functions, more space than what its real weight within our inquiry warrants.

The task we wish to solve is a purely theoretical one, without becoming a sheer logical exercise. Cataloguing theory has, naturally, always to remain the servant

[6] The reasons for our not having to examine this particular question will be given later on; see pp. 71–72.

of practice. Accordingly, we shall draw now and again practical inferences from our theoretical findings.

Finally, by way of warning, I want to emphasize that I shall write only about facts which most cataloguers are not, or only partly, unaware of, the whole importance of which, however, they do not often realize. In consequence of this, they frequently omit to integrate them into a system, to draw even the most cogent inferences from them, and to co-ordinate their theoretical propositions, formulae, rules, and practices with them. The explanation for such astonishing failures is to be sought, I think, in the Aristotelian truth, experienced by most of us at our own expense often enough, that nothing is as difficult as to discover the obvious.

5. THE SUBJECT-MATTER OF THIS STUDY IS NEITHER OBSOLETE NOR UNTIMELY. Some librarians may think the theme of this study hopelessly out of date. After all, it cannot be denied that today the attention of librarians is focused chiefly on the subject approach to the materials they are entrusted to administer. An outsider, browsing at random in library literature and finding that it is only exceptionally that author-title cataloguing is even mentioned, is liable to get the impression that this branch of cataloguing must have lost all its former importance, and that to spend valuable time in investigating its principles and facts must be, therefore, a plain mistake.

It is, however, he who would be mistaken. The author-title catalogue is surely still the primary catalogue in most libraries. Even in our days it is the author-title access, and not the subject access, which serves the majority of patrons most satisfactorily, especially in large research libraries, with which alone we are here concerned.[7] In this respect no substantial change is to be expected within a reasonable time – seeing that the demand for adequate author-title access is rooted in the most cardinal needs which these libraries are obliged to satisfy. In spite of the rapidly growing impetuosity of the claim for increasingly better and more intricate subject access, we must not forget that even today a very considerable proportion of patrons do visit the library not for subject information, but simply for getting a particular book or work. Things will have to change a great deal before a dwindling of this class of patrons and a resulting superannuation of the author-title catalogue can ensue.

But the objection of untimeliness might be raised against our choice of subject also on a narrower and at a first glance more plausible basis. One might argue that this subject has never been so inopportune as today, seeing that the card catalogue may before long be superseded by the computer and that this would involve considerable changes in the functions of descriptive cataloguing in the broader sense of the word. There are even authors who do not think it impossible that the author-title approach may come to be discarded altogether. "The extreme question must be considered," – says C. D. Gull – "will rules for author and title entries be required at all for electronic information systems?"[8] This question must be answered decisively in the affirmative. The question itself testifies an absolute lack

[7] See § 6 and Appendix I.

[8] GULL, C. D.: How will Electronic Information Systems Affect Cataloging Rules? *Library Resources and Technical Services*. Spring 1961, p. 139.

of comprehension of the nature and scope, not only of the rules, but also of the entire role of author–title cataloguing. Besides, Gull also forgot that even in an era of automation printed and written catalogues and book-lists will still continue to play a very important role.

Now, it is, of course, not at all impossible that automation may generate brand-new functions of author-title cataloguing, by providing access to books, e.g., by key title words, the publisher's name, the date or place of publication, etc. But all the possibilities of the appearance of new functions do not provide any excuse for postponing a close examination of the present functions of cataloguing, due already for a long time past. One would have some justification for considering a postponement, if there were reason to assume that one or the other of the functions would be discarded or considerably transformed by automation. There is, however, no reason for such an assumption. The development which has fashioned the tasks the catalogue is actually performing was, evidently, determined by the comparative urgencies of the demands of patrons and librarians; present-day functions must have had their origin in the most urgent among these demands. It would be absurd to suppose that the automated catalogue, while performing new functions that are of comparatively minor importance, should at the same time mutilate, or even jettison, one or the other of the most valuable old services which the author-title approach is able to perform.

So, I think, we may infer without hesitation that the prospect of automation increases the importance and timeliness of an investigation of the functions and objects of author-title cataloguing instead of diminishing them. It is liable to call much more imperiously for precision in the specification of these functions and objects than traditional cataloguing ever did.

In one particular respect, however, automation is very likely to bring a decisive change into descriptive cataloguing: by the dethronement the main entry will probably suffer in automated catalogues. This is, however, a matter quite irrelevant to us for two different reasons. First, in non-automated catalogues the main entry will still preserve its exalted position; and second, the choice from the three different forms of entry, and the whole question of the primacy of the main entry too, is simply irrelevant to the subject of this inquiry. We shall explain the reasons for this later on.[9]

To sum up: the advent of automation in cataloguing cannot be turned into an argument for the untimeliness of our investigation.

6. OUR INVESTIGATION OF THE OBJECTS OF CATALOGUING WILL BE CONFINED TO THE AUTONOMOUS OBJECTS OF STANDARD CATALOGUING. We do not intend to enter into every detail of the subject as designated in the title of this book. To rid our train of thought of less interesting detail, we shall exclude two particular aspects of this subject. We must delimit these aspects with the greatest possible precision, first of all for the simple negative reason that it is logically indispensable to be minutely explicit about the omissions we are intending to make; and secondly, because our outlining these omissions with clarity will also yield a positive result: by specifying the fields and problems we do not intend to deal with, we shall get, into the bar-

[9] See pp. 71–72.

gain, a glimpse of all the principal kinds of tasks and objects of cataloguing by the examination of which the investigation of the subject designated in the title of this study might be brought, even in matters of minor interest, to completion.

All the objects about which author–title cataloguing conveys information may be divided in the first instance into two main classes. To the first class those belong which get separate entries of their own in the catalogue, under headings derived from their own distinguishing marks. So in this class the objects themselves fix their own place in the catalogue, or more precisely, the place of the units of information referring to them. We propose, for this reason, to call them 'autonomous objects'.

The other class of the objects of cataloguing do not get any separate entries of their own. A treatise forming a part of the contents of a one-volume collection, for instance, which, although mentioned in the body of the entry for the collection, is given no added entry under its own author and/or title, is an object about which a certain sort of information is conveyed, but which is denied the status of an autonomous object, there being made no entry directly for serving its retrieval. So the information actually given about that treatise serves, obviously, merely the purpose of enriching the information about the autonomous object, the collection, in which it is contained. The objects belonging to this class we propose, therefore, to call 'subordinate objects'.

This particular division of the objects of cataloguing into two sharply separated classes corresponds to the division of the entire domain of author-title cataloguing into two distinct provinces. One of these is concerned with the choice and form of the heading and entry-word of catalogue entries, the other with descriptive cataloguing in the narrower sense of the word. The first province deals with, and its methods and rules are determined by, the autonomous objects of cataloguing, whereas the other is concerned with, and its methods are fashioned according to, the subordinate objects. The efficiency of the author-title catalogue depends primarily on the adequate solution of the problems emerging in the first province, so these are more important, and also far more vividly contested. They divide in their turn again into two classes; those belonging to the first are concerned with the choice of the heading, while those of the second relate to the form of the heading and the choice of the entry-word.

Our investigation will be focused only on the first section of the first province. We shall have nothing to do with questions of descriptive cataloguing in the narrower sense, nor with those regarding the form of the heading. Consequently, our investigation will not go beyond a specification and classification of the autonomous objects of cataloguing, nor beyond a scrutiny of its functions focused on such objects. Therefore, for brevity, in the following we shall use the term 'objects of cataloguing' to denote the 'autonomous objects of cataloguing'. This is, then, the first restriction we are going to impose on the following analysis.

The border-line between autonomous and subordinate objects is, however, in several respects liquid. First of all, in different fields of special cataloguing there occur several kinds of autonomous objects which in the sphere of standard cataloguing do not attain that elevated status.

The term 'standard cataloguing' is used here to designate that branch of author-title cataloguing which is conducted according to those rules for cataloguing books and book-like materials which form the main body of rules in most national codes designed for the use of large general and research libraries. As opposed to standard cataloguing, the term 'special cataloguing' is applied, first, to the cataloguing of those classes of printed documents which either use a medium other than written language (printed music, maps, and pictorial representations, e.g., reproductions of works of art) as their main medium of communication, or to which a greater or lesser than ordinary importance is attributed, and therefore a more or less minute than the standard treatment is applied (incunabula, ephemera, etc.); and secondly, to the cataloguing of other than printed documents (manuscripts, gramophone records, photographs, motion pictures, filmstrips, etc.).[10]

As an example of the objects the status of which changes with the varying fields of cataloguing they happen to belong to, we may mention a printer's production. In cataloguing an incunabulum many libraries devote to the printer an entry of his own, and turn it thus also in its capacity as a particular printer's product into an autonomous object, or more precisely, into one of the components of an autonomous object. In standard cataloguing, on the other hand, in this particular capacity books always remain mere subordinate objects of cataloguing.

But even in the domain of standard cataloguing itself there do occur numerous objects with a variable status. The treatise making a part of a one-volume collection is in our example mentioned above reduced to the status of a subordinate object. But if a library chooses to assign to it an added entry under its own author and/or title, then it is turned by the catalogue of this library into an autonomous object. This example shows that many kinds of subordinate objects – theoretically even all of them – are potential autonomous objects; their character depends solely on the treatment the code and the cataloguer choose actually to apply to them.

Those kinds of objects which are alternately autonomous and subordinate, may be divided thus into two classes. The first class comprises those which never become autonomous except in special spheres of cataloguing, while the second one contains those which may do so also in standard cataloguing. This distinction is the basis of the second principal limitation we shall impose upon the following inquiry: we also exclude all those objects which do not become autonomous except in fields of special cataloguing but never in standard cataloguing. This class of objects is divided again into two sub-classes. A large proportion of the objects belonging to it is not unknown to standard cataloguing, but played here – at least up to now – always the role of subordinate objects only. The rest consists of objects alien to standard cataloguing even in the capacity of subordinate objects. The exclusion of these two sub-classes of objects from our inquiry implies naturally an exclusion of all those functions, too, which are confined to the recording of such objects only.

The motive for imposing this restriction on our inquiry is, again, the wish to reduce its length. Special materials are apt to display specific features and

[10] Here we have to make a special mention of two intermediate types of library materials: of the photocopies and microfilms of such printed documents as are subject to standard cataloguing. Being strikingly different from printed documents technologically, these two ought by rights to be classified as special library materials. From our point of view, however, there is no reason for distinguishing them from the objects of standard cataloguing, seeing that their entries differ neither in heading nor in body from the entries of the reproduced printed documents. In consequence, everything we shall in the following say applies also, without any qualification, to this class of photocopies and microfilms.

peculiarities relevant to cataloguing, and thus they also incline to generate objects and even to call for functions of cataloguing unknown in the domain of standard cataloguing. We do not want to encumber this inquiry with all these petty details, although strictly spoken they do belong to the subject that is designated in the title of this book; because we wish to confine our attention only to such aspects of this subject which are of general importance. The functions and autonomous objects occurring only in fields of special cataloguing do not belong to these aspects.

There is, however, one point which we had better mention in this connection by way of a digression. The narrowing down of the scope of our inquiry to an investigation of standard cataloguing only might appear to be subject to the objection that cataloguing ought to be uniform for all library materials, and consequently cataloguing theory, too, must be unique, unvarying and valid for all sorts of materials. If this were true, the proposed second limitation on the subject of this study would be entirely unwarrantable, seeing that if the theoretical argument to be developed here applied only to a single sector of the whole cataloguing area, one would have to build up for the other sectors one or more further theories different from the former. A thorough analysis of this objection would, however, much retard the progress of our argument proper. So we propose to relegate it to Appendix I, and in the present context to confine ourselves to stating simply its results, which are the following.

Inevitably, theory must investigate the functions and objects of author-title cataloguing separately for each sector of cataloguing. Consequently and besides, owing to the fact that standard cataloguing is by far the most important among these sectors, our intention of confining our study to the latter is not only free from the objection indicated above, but on the contrary, it is to be approved of, being a happy device with the help of which the double-barrelled advantage of embracing all the facts and problems of general concern, and of excluding at the same time every comparatively unimportant and uninteresting detail, can be secured.

7. OCCASIONALLY WE SHALL TRANSGRESS THE LIMITS OF OUR SUBJECT-PROPER. In the foregoing paragraph we have specified those aspects of the theme of the functions and objects of author-title cataloguing which will not be dealt with in this study. Now we must add that occasionally we shall also transgress the strict limits of the theme designated in its title. Now and again we shall be obliged to touch upon themes belonging not to our, but to another chapter of cataloguing theory. If we could fall back on an established and elaborate theory of cataloguing, in such cases perfunctory references would suffice. In the absence of such a theory, however, in several instances we shall be compelled to insert into our train of thought – in the form of digressions retarding the progress of our argument proper – a minute examination of concepts or problems which could otherwise have been dispensed with. This is, on the one hand, rather cumbersome; but it provides us, on the other, with opportunity for touching upon a few neglected questions of cataloguing theory not immediately pertaining to our subject, and broadens, thus, the scope of this study.

CHAPTER II

THE MEANING OF THE TERMS 'FUNCTION' AND 'OBJECT' OF AUTHOR-TITLE CATALOGUING

8. BOTH TERMS HAVE SEVERAL DIFFERENT DENOTATIONS RELEVANT TO OUR SUBJECT. Together with quite a lot of other terms of author-title cataloguing, the terms 'function' and 'object' are used in several different senses, often overlapping each other. We must inspect these in order to be able to draw a precise line of division between the particular meanings which we propose to attach to the two terms, respectively, and between their other meanings. Let us scrutinize first the term 'function of author-title cataloguing'.

If used in the sense relevant to our subject, the word 'function' is in common speech approximately synonymous with 'task', 'duty', 'service'; denoting, however, in more discriminative usage only basic and comprehensive tasks and services. As a catalogical term, too, the word is used to denote only tasks which have the most general and comprehensive character and the widest scope among all tasks cataloguing is or may be fulfilling. The objective of any investigation of the theoretical theme usually labelled as 'the functions of cataloguing' is to illuminate the essence of the catalogue, that is, the catalogue being a tool, the principal services it renders. The conciseness and clarity of the results of an investigation of this theme depend on the condition that the essential and fundamental services of the catalogue should be sharply distinguished from those belonging to lower orders, which implies that they must also be called by a distinctive name of their own. To delimit accurately the sense in which the term 'function' is to be used in the theory of author-title cataloguing is, therefore, not only a matter of logical tidiness, but also one of practical importance. Although this seems self-evident to me, I have some misgivings that the following analysis of this topic might be considered superfluously lengthy and meticulous. That is why I recommend to scan the pages devoted to this subject in Benoyendra Sengupta's book on *Cataloguing* published in 1964.[11] Sengupta's enumeration of three objectives, two basic functions, eight other functions and three services of library cataloguing succeeds in producing a hopeless muddle, and proves convincingly that a thorough examination of the logical foundation upon which a serviceable meaning of the term 'functions of cataloguing' can be based is still very far from being superfluous.

[11] SENGUPTA, BENOYENDRA: *Cataloguing. Its Theory and Practice.* Calcutta, The World Press, 1964, pp. 4–8.

9. THE MOST COMPREHENSIVE DENOTATION ONE MAY GIVE THE TERM 'FUNCTION'. Even after it has been made clear that the term 'function' is to be applied only to the most fundamental and comprehensive tasks of the catalogue, the denotation of this term is still very far from being unequivocally fixed.

The logically first, i. e., most general and comprehensive definition of the term 'function of library cataloguing' is: to convey information about the library holdings, or, since these holdings are composed of a mass of distinct individual items, about all these items. Owing to the fact that it has an extremely general character and does not disclose anything either about the nature of the objects of the information the catalogue contains, or about the nature of the needs it can satisfy, this definition fits not only author-title, but also all other branches of cataloguing. By appending to it various qualifications concerning the peculiar character of the information conveyed about an item in the holdings, the difinition can, however, be given diverse forms, each of which will be confined to fitting merely one kind of library catalogue. The structure of each kind of catalogue is, namely, based on only one or a few aspects of the books and other documents about which it is conveying information. Each has its own peculiar point or points of view, as well as its own peculiar means. Each organizes the information it contains on the grounds of those aspects and characteristics of the documents which are discernible from its particular point(s) of view. The subject catalogues, classified as well as alphabetical, for instance, provide an approach to documents directly by their subject contents. The structure of catalogues of this kind is based on formulae or symbols expressive of these contents. There are special catalogues the structure of which is based on the year and place of publication of the documents, or on the names of their illustrators, or of the persons the portraits of whom they contain, etc.; not to mention the various hybrid types of catalogues which combine the viewpoints of two or more pure types.

The author-title catalogue arranges the information by the tags or signs attached to each particular book and work in order to distinguish it from other books or works, respectively. The structure of this catalogue rests on these tags, these distinguishing marks of a formal nature, which we shall inspect at some length presently. This gives to the author-title catalogue an essentially formal character,[12] which contrasts it sharply with the subject catalogues, the other principal class of library catalogues. Therefore, by adding to the above general definition of the 'catalogue-function' the qualification that the information conveyed is arranged by the *formal distinguishing marks* of the documents, it can be turned into a distinctive definition which fits only the author-title catalogue. This qualified definition is, however, still in a high degree formal and does not suggest any idea of the concrete nature of the services rendered by the catalogue defined this way. This is the reason why the possibility of this definition is always overlooked, although it constitutes the only one of presenting the function of author-title cataloguing as a single and homogeneous service, and at the same time also the possibility of attaching to this term the most comprehensive meaning possible. That is why theorists keep omitting to take this logically first step, and insist on approaching the subject only on a one degree lower level of abstraction and generalization, by applying the designation 'function' to several services of the author-title catalogue each of which constitutes only a section of its whole service.

[12] This topic will be taken up again at greater length in other contexts.

10. THE DENOTATION WHICH THE TERM 'FUNCTION' IS GENERALLY GIVEN IN OUR PROFESSIONAL LANGUAGE. Crudely defined, these 'second-level functions', usually called simply *the* functions of the author-title catalogue, are considered to consist in communicating information about the books, the works, and the authors' 'oeuvres', contained in the library. This denotation of the term 'functions' manages already to convey a notion of the actual nature of the author-title catalogue and its services. It indicates in a comparatively more concrete, but still highly general manner the objects in a communication of information about which the respective essence of every function consists. Since the logically first denotation of the term 'function of the author-title catalogue', developed in the foregoing paragraph, is, as we have seen, not particularly revealing, we need not concern ourselves any longer with it. So there is no reason why we should not adopt the generally accepted denotation of this term; in what follows we shall use the word 'function' – without any qualification – to denote these 'second-level' tasks.

This means first of all that the upper limit of this denotation excludes the comprehensive single function as defined in the foregoing paragraph. Its lower limit excludes, on the other hand, all those tasks and services of the catalogue which belong to still lower orders, still lower levels of abstraction and comprehensiveness. To distinguish sharply between these latter tasks and between those for which we have reserved the designation 'function', i. e., to draw the latter's lower boundary line, this is our principal task as regards the functions of the author-title catalogue. True, the generally accepted denotation of the term 'functions of author-title cataloguing' does imply some sort of answer to the question of the location of this boundary line, but only an approximate and rather a vague one. The task before us is to complete now this answer, to make it more exact by filling in its gaps as far as possible. The importance of this task must not be underestimated. Vague notions on the classification of relevant facts are, almost to the same extent as vague concepts, liable to entail false inferences and mistaken actions. Let us illustrate by a few examples the significance of drawing the lower limit of the class of tasks to be called 'functions', with the greatest possible clarity.

To offer descriptions which enable the users easily to identify each particular book and to distinguish it from every other book, belongs undoubtedly to the primary duties of the catalogue. So does also the endeavour to satisfy, within reasonable limits, even those readers whose information on the particular book they are looking for is deficient. Another important duty of the catalogue is, indubitably, to facilitate the recognition of a particular edition of a given work by indicating (among others) its year of publication; or else, to make a book retrievable also under a short title figuring on its cover and differing considerably from the title-page title, etc. Questioned about their opinion on these tasks, a great majority of librarians will certainly agree in recognizing them as tasks that the catalogue must not fail to perform. There is, so far, nothing wrong with this answer. But, if asked for his opinion on the tasks of the catalogue, a cataloguer answered by enumerating, along with the three functions formerly referred to, also the four tasks here mentioned, this answer would not be tenable any more. The extent to which these tasks differ in importance forbids mentioning them in the same breath, within the same specification, as if they were things of the same order. Giving such an answer, our fictitious librarian would be, evidently, deceived by the linguistic fact that the words 'function', 'task' and 'duty' are commonly being applied indiscriminately to a broad scale of things which belong, in consequence of the great differences in their importance, to distinctly different orders, to qualitatively different categories. The duty to disclose the year of publication of the books recorded, is only one among the many means employed in securing the performance of duties of considerably higher order: of the duty to facilitate the quick identification of books, and of the duty to enable the reader to find out that particular one among the editions of a work which suits his needs best. But the two latter tasks are in their turn also mere means of rendering the catalogue fit to fulfil its 'functions' in the above intimated sense of the word. It would heavily impair the lucidity of any representation of the objectives and services of the catalogue if it ranked these three layers of tasks in the same category. One would make the same mistake by disregarding those differences of order and category which establish definite lines of division between the duty of recording striking cover-titles, the duty of endeavouring to help readers deficiently informed, and the duty of disclosing whether or not a particular book is contained in the holdings (i. e., of discharging the first function).

It is self-evident that we must not apply the denomination 'function' to a task shouldered by cataloguers only as a means of promoting the fulfilment of a task of higher order, nor to a task which is simply a section, a component part of a more comprehensive task with a broader and more inclusive scope. It is so, because both these kinds of tasks obviously belong to lower grades of the hierarchy of tasks than those for which we intend to reserve this denomination. This is a fundamental precept underlying our classification of the tasks of the catalogue. It is, however, to be supplemented by stressing that some tasks – notwithstanding the fact that they cannot be subsumed under a function, either in the capacity of an instrument of the fulfilment, or in that of a component part of the latter – are not to be ranked with the functions, simply in consequence of their comparative insignificance.

11. THE CURRENT FORMULAE OF THE DEFINITIONS OF THE THREE FUNCTIONS. In the previous paragraph we have drawn the limits of the denotation we propose to assign to the term 'function', i. e., we have delimited this denotation negatively. We have now to fill up this formal frame, by equipping the term with a positive and concrete denotation.

Let us start from two facts we have already stated. First, that author-title cataloguing organizes the information about its objects by means of formal characteristics, by formal distinguishing marks of these objects. Secondly, that this inquiry is not concerned with objects not retrievable in the catalogue under their own formal distinguishing marks, with objects that are not autonomous. In consequence, we shall not concern ourselves with those formal characteristics of the objects of cataloguing, not even with those of the autonomous objects, which do not appear in the heading of an entry, and do not play a part in securing access to an object, i. e., which do not belong to the agents that can make objects autonomous. Of course, it would be possible to make documents accessible entering them in the catalogue under any formal mark one chooses to include in their entries; for instance a book might be made retrievable in its capacity as a product of the Macmillan Company, or as a volume containing 269 pages, etc. Cataloguers, however, do not make use of many of these possibilities. Many kinds of formal marks appearing in entries remain excluded from the ranks of those marks by which documents are being made retrievable. They are not traceable in the catalogue, and so are, in consequence, also the documents to which they are attached, at least in their capacity as objects invested with them. Many potential autonomous objects, and also potential functions of cataloguing are, thus, simply discarded and ignored.

What justifies the cataloguer to perform one possible task, and to discard the other? What are the appropriate viewpoints from which he may start when undertaking such a selection?

Of course, this question cannot be answered on a purely logical basis, seeing that it is impossible to conceive of an abstractly perfect author-title catalogue. Cataloguing being a practical art, we have to ponder the optimality of the catalogue and not its absolute perfection. This can be reached – in the given historical situation – by a sound balancing of the actual utilities of all its obtainable services with their respective costs, on the one hand, and with the available economic resources, on the other. At the first glance even this task seems inaccessible to theoretical solution, as not only the economic resources, but also the demands of patrons and the utility of the services, do vary considerably, not only in the different countries, but also in individual libraries. In spite of this, however, the great majority of present-day librarians seem to agree upon the functions, the fundamental tasks which a satisfactory author-title catalogue ought to perform.

At the Paris International Conference on Cataloguing Principles in 1961, it became clear that, notwithstanding the fact that actual practices do often fall short of this standard, the majority of expert cataloguers of the world judge the performance of three functions – at least in large general and research libraries – to be indispensable to the efficiency of the author-title catalogue. These functions are, as this was already intimated above, to convey easily and quickly retrievable information about (*i*) each particular book, (*ii*) all the editions of each work, and (*iii*) all the works of each author, which are available in the library.

This statement slightly differs in its form not only from the formula given in § 2 of the *Paris Statement of Principles*,[13] but also from most of the different formulae advanced by individual authors before and since Paris. Its content is, nevertheless, in all essentials congruous with the contents of most of these formulae. So we may regard it with good reason as an appropriate expression of the opinion prevailing among contemporary cataloguers, and refer to it, for brevity, as the 'current formula' of defining the functions of the author-title catalogue. The deviations of this formula from the Paris one we shall duly account for in the proper contexts later on.

In the following inquiry we are going to use this formula as its point of departure. We shall examine first of all whether the statement, that author-title cataloguing performs these three, and only these three, separate and distinct functions, is or is not verified by present-day actual conditions. This examination should consist of a positive and a negative branch. We have to scrutinize not only

[13] *Statement of Principles, Adopted by the International Conference on Cataloguing Principles, Paris, Oct. 1961.* Annotated edition with commentary and examples by Eva Verona, assisted by F. G. Kaltwasser, P. R. Lewis, Roger Pierrot. London, IFLA Committee on Cataloguing, 1971, p. 6.

(*i*) whether the three tasks mentioned in our formula belong really to that order of tasks which we have resolved to denominate as functions but also (*ii*) whether it is not possible to discern any other task at present actually performed by author-title catalogues, or reasonably reclaimed by readers, which could be classified as belonging to them.

The first question could be answered in the affirmative in a logically satisfactory manner by demonstrating the two facts that none of the three tasks usually called functions is subordinate to any other task, to another function, and that each of them is really of fundamental importance. All this we shall demonstrate in appropriate context in the course of the following argument.

12. CRITICISM OF TWO ATTEMPTS AT RECOGNIZING A FOURTH FUNCTION. An answer logically satisfactory enough to the second of the above specified questions could not be obtained except by a systematic examination of all the tasks the author-title catalogue is in fact performing. Such an examination would be, however, so long-winded and tiresome, and promises so little result, that it does not seem practicable at all. So the possibility of an error, of our overlooking a plausible candidate for the rank of a function, cannot be entirely excluded. One meets now and again with an attempt at ranking a further service with the three functions, but these attempts prove on closer inspection regularly unsuccessful. This is, I think, a reasonably reassuring argument in favour of a negative answer to our second question: apart from its three acknowledged functions, the catalogue does not perform any task that could be classed as belonging to that order for which the designation 'function' has been reserved. By way of illustration, I shall now review two of these attempts, both of which have already been given up by their originators although at first glance they might be considered well-founded for apparently weighty reasons.

The first of these attempts has been made – an uncomfortable fact! – by myself. In an earlier paper I objected to § 2 of the *Paris Statement of Principles* omitting to rank with the functions a rather universally accepted and very important service of the catalogue: its efforts within reasonable limits to help catalogue-users inadequately informed either about the author and/or the title of a book or work, or else about the rules of cataloguing.[14] I raised this objection on the consideration that this service essentially differed from all services indubitably subordinate to the functions, i. e., from all tasks evidently *not* to be ranked as 'functions'. Such tasks, for instance, describing books in a manner which facilitates their easy and quick identification, or making every book traceable also for those users who are acquainted only with that form of the title or of the author's name which figures in the book, I reasoned, were mere means, though indispensable ones, of the fulfilment of one or of several functions. In contrast to them, the help offered to the insufficiently informed catalogue-user is not necessarily included in the three functions, it is no essential condition of their fulfilment, and at the same time it is a service the fundamental importance of which is clearly reflected in its general recog-

[14] DOMANOVSZKY, Á.: Der Code-Entwurf der Pariser Katalogisierungskonferenz. *Libri*. 1962, No. 3, p. 213.

nition. When I made my critical remark, I thought that these two facts sufficed to justify classing this service as a function. Erroneously, I now think.

Between the above-mentioned obviously subordinate two tasks and the service in question there is at the most a difference of degree, a difference in their comparative significance; but if there is such a quantitative difference at all, it is certainly not great enough to turn into a qualitative one, and so to justify a different classification of these two kinds of tasks. It is not entirely impossible to perform the three functions without performing at the same time also the two subordinate services mentioned above as examples; it is only impossible to perform them as efficiently as when the latter are being fulfilled as well. The same applies, however, also to the task of helping the inadequately informed catalogue-user.

Moreover, at present I think it already obvious that a task which has no peculiar object or objects of its own, cannot be ranked with the three functions. The task under consideration lacks this criterion; it consists merely in providing additional access points to the objects embraced by the three functions, but does not add any new items to the aggregate of these objects. It does not augment the range of information conveyed by the catalogue, only the efficiency with which the information it already contains is conveyed. Being thus a mere means of rendering the performance of the functions more efficient, it is a task obviously subordinate to the functions, in spite of the fact that it is not necessarily, logically, implied in their very concepts. Among the generally accepted and undeniably important tasks of author-title cataloguing there are several others of a similar character. Our earlier propounded but now withdrawn argument in favour of exalting the task under consideration to the rank of a function, if it were valid, would apply to all these tasks too.

This shows, by the way, that all the subordinate tasks the author-title catalogue of our days actually performs are not necessarily implied in the concepts of its three functions, some might be jettisoned without preventing the performance of these functions, though usually not without impairing the efficiency of this performance. Correspondingly, the current formula of defining the functions does not determine, nor even hint at, the actual or the optimal measure either in which these functions are, or ought to be, performed.

We must warn against considering this a deficiency of the current formula. The latter has as to articulateness and precision several shortcomings which it is not at all impossible to eliminate, but the feature under consideration does not belong among these. It is simply a consequence of the fact that the author-title catalogue is a complicated structure which cannot be characterized exhaustively by a concise formula. The niceties, the less important details, of its characterization must be postponed, and must not be mixed up with basic matters. This is a principle of the technique of thinking; and we might mention in passing that it is a principle to be observed with the utmost rigidity first of all by code-makers. Many cataloguing codes today in application exhibit illustrations of a mixing-up of fundamental matters with insignificant ones, among them even downright comical cases, for instance, orthographical or typographical instructions popping up astoundingly in their very first paragraphs, mixed up with the most cardinal rules for the choice of the heading. Such blunders, extremely damaging to the lucidity as well as the instructiveness of a code, are signs of the codifier's failure either to discern the logical hierarchy of the multifarious tasks author-title cataloguing comprises, or to act upon this discernment.

Another proposal for transferring a fourth item into the class of functions had been made by Eva Verona in the first draft of her code compiled for the Federation of the Library Associations of Yugoslavia. In her specification of the fundamental duties of the catalogue, Verona had also mentioned, along with its three traditional functions, the task that "it has to convey information about the single volumes belonging to a series of books, to a collection, or to collected works published in several volumes; that is to say, about those single volumes of the publications specified which bear apart from the separate, specific titles of the single volumes also a common title that turns all of them into a single entity".[15]

As to the series, the primary task of author-title cataloguing is generally held to consist in providing access, under their respective sets of formal marks, to the single volumes which constitute them. This is, however, a task included already in the first three functions, which, by the way, are defined by Verona entirely in accordance with the current formula. The specific circumstance the cataloguer has to deal with under the conditions in which according to Verona a fourth function originate – the existence of items which are not only separate single books, but also part of a more comprehensive, complex publication – calls, indeed, for a specific additional service. This service, however, does not consist in entering such volumes as separate items, but in entering them additionally also under the distinguishing marks common with all the separate parts of the complex publication. This task, however, cannot be considered a separate and distinct function either, because the complex publications under consideration, the series, are to be subsumed under the concept 'book' of the current formula, and so the task Verona is referring to is already included in the three functions. What we are really confronted with here, is merely the fact that in a single physical item there are often two or more distinct catalogical items (e. g., a separate monograph, and a constituent part of a series) amalgamated, each of which may be looked for by catalogue-users, and must be recorded therefore in a manner appropriate to it simply on behalf of the three functions. With the phenomenon, that one 'book' in the everyday sense of the word may unite two or more distinct 'books' in a catalogical sense both of which must be recorded separately, we shall deal at length later on.[16]

The phrase "izdanja sabranih djela u vise svezaka" refers, I believe, to collected works by single authors published in several volumes. The task of dealing with this kind of complex publication cannot be ranked as a new function, either, though from our present point of view such collected works are directly opposed to series. As to the collected works of an author, the primary task consists, obviously, in securing access to the complex publication, not to the single volumes that constitute it; but the task of making these single volumes traceable also under

[15] Since I am not quite confident of the correctness of the above translation, I think it necessary to quote here the original text too: "Abecedni katalog mora obavijestiti o pojedinim svescima nizova knjiga, skupnih publikacija i izdanja sabranih djela u više svezaka, tj. o svescima onih publikacija kod kojih pojedini svesci pored svojih posebnih, specifičnih naslova imaju još i zajednički skupni naslov koji ih povezuje u jednu cjelinu." *Pravilnik za Izradbu Abecednih Kataloga. Nacrt.* Izradila Eva Verona. Snopič 1. Beograd. Savez Društava Bibliotekara Jugoslavije. 1962, p. 6.

[16] See § 28.

their own titles is only of secondary importance, which is neglected not only by printed, but also by many other library catalogues, at least in large general and research libraries, without any notable harm.

We must not omit, however, to emphasize that her having explicitly stated that all the publications under consideration should be recorded consistently in two different capacities, as two diverse objects of cataloguing, is, indeed, a great merit of Verona, seeing that codes habitually fail to be articulate on this point. It is only her ranking this service as a separate function with which we cannot agree.

13. DO THE FUNCTIONS THE CATALOGUE AT PRESENT PERFORMS SATISFY ALL ACTUAL DEMANDS? The foregoing paragraph was devoted to the question whether or not there exist among the tasks which present-day author-title catalogues actually perform, such that the current formula does not mention in spite of their belonging to the tasks we have classified as functions. In order to satisfy the requirements of logic, we should, now, also ask whether advanced practice and theory do or do not fit present-day conditions. That is to say, we should examine, first, whether one or the other of the three functions did not become already obsolete, and second, whether, reversedly, practice as well as theory have not failed to discern a task of the order of the functions the performance of which would be fully warranted under the conditions by now already reached.

The first question in this second pair of questions is only a modified form of the first question raised in § 11, and will be answered together with the latter in the course of the following detailed analysis of the functions. The second one we can quickly dispose of. It does not seem probable that in the present state of library technique author-title cataloguing would be capable of performing, at a cost justified by the satisfaction catalogue-users would derive from it, a new, hitherto not-dreamt-of task fit to be ranked with the functions. At any rate, I have not imagination enough to think of such a task. Changing conditions may, naturally, generate brand-new functions of descriptive cataloguing in the broader sense of the word. An automated catalogue, for instance, may show up the books of a library according to place or year of publication. But until the great majority of catalogues are non-automated, the conveying of these two classes of information cannot become generally adopted functions of descriptive cataloguing, since their costs are liable to be out of all proportion to their utility.

14. THE RELEVANT DENOTATIONS OF THE TERM 'OBJECTS OF AUTHOR-TITLE CATALOGUING'. As already mentioned, the three propositions contained in the current formula about the catalogue-functions indicate merely the essence of these functions, and even this only in an excessively concise form which is sometimes directly inaccurate and misleading. It needs laborious and long-drawn investigations to disclose the complexity and intricacy of the real facts hidden behind these simplifications; and just these investigations involve us in a minute inspection and classification of all the objects a communication of information about which constitutes the three functions. The concept 'object of author-title cataloguing' becomes thus one of the crucial concepts of this inquiry, and so it is very important carefully to distinguish between the diverse possible denotations which one may attach to the

corresponding term, and clearly to fix with which one among these we shall invest it.

The most general definition of the concept 'object of author-title cataloguing' is, obviously, that which corresponds to the most general definition of the concept 'function'. The term corresponding to the concept so defined denotes the holdings of a library, conceived as a unity, a oneness; in so far as the specific methods of author-title cataloguing are applied to it.

The next move is, here again, to proceed from the definition which presents the object of cataloguing as a single, undifferentiated, monolithic unity, to one which is already focused on the items, the particles, of which this monolithic object is composed. There are several different possibilities of dissolving that unique object into its constituting units. One among these is the substitution for the unity 'holdings' of the sum of all the detached books and book-like documents, subject to standard cataloguing, which the holdings contain. At a first glance this appears to be the most simple and plausible way of solving the task in hand. This impression is, however, deceptive, the solution is inadequate for two different reasons. We shall see later on that there are some kinds of objects of author-title cataloguing which are neither books, nor documents of any other kind. And what is even more important, this concept of the objects of cataloguing is still in an unsatisfactorily high degree formal, and discloses, in consequence of its representing these objects as an undifferentiated homogeneous mass, only very little about the concrete services, the real essence, of the catalogue. The concept 'object of cataloguing', and the denotation of the corresponding term we are in need of, have to be made considerably more concrete and differentiated, revealing all the important manifold differences, first, between the diverse types into which at a closer inspection the apparently homogeneous total mass of individual objects disintegrates, and secondly, between the tasks the cataloguing of these different types of objects comprises.

15. THE SECOND DENOTATION OF THE TERM 'OBJECT'. 'BOOKS', 'WORKS' AND 'AUTHORS' OEUVRES'. The first step towards the disclosure of such a concrete and complex denotation of the term 'objects of author-title cataloguing' we must take, obviously, by setting off from the triplicity of the functions of the catalogue. It is not possible to reach a satisfactory answer to our question if we confine ourselves to taking books into consideration only in their capacity as distinct and detached physical things. One has to take account of them also in their capacity as members of two different kinds of families of books: one formed by all the materializations of a particular work, while the other, by all the materializations of the works by a particular author. The single objects of cataloguing are, thus, not simply all the single books contained in the holdings, but three diverse aspects of all of them.

Perhaps we had better examine here, by way of a digression, the following question. Our aim being to gain a *complete* inventory of all the objects of standard cataloguing, is it not a mistake to confine our attention to only three aspects of books: to those revealed by their inspection from the points of view of the three catalogue-functions? After all, along with these, books have several other aspects, many of which are also regularly recorded by catalogues. Indeed, every fact about which a catalogue entry conveys information illuminates a new aspect of the book described. Would it not be more appropriate theoretically, or more expedient practically, to look upon all–instead of three – of these aspects, upon all the different sides of books the catalogue does throw a light on, in one word, on the entire range of the facts it discloses, as upon the constituent units, the particles, which the monolithic object of cataloguing, the 'holdings', are composed of?

To determine in this way that second meaning of the term 'object of cataloguing' we are now in search of would be, though not objectionable logically, of little use. That meaning would be still too formal in character, and fail to throw any light upon the nature of the services of the catalogue and on the needs these services may satisfy.

The catalogue-user is concerned directly with the documents themselves; the aspects of these documents which the catalogue reveals, and the facts it discloses about them, are, as the mere means of obtaining his real end, of only subordinate interest to him. Moreover, both these conceptions of the objects of cataloguing would include, along with the autonomous, also the subordinate objects, with which this inquiry, as we know, is not concerned.

We had better drop, then, both these conceptions, and revert to the triplicity of objects implied in the triplicity of the functions of the catalogue, that is, to dividing the total mass of objects, first of all, into the three main classes correlated with the three functions. These three classes of objects may be labelled shortly and crudely as 'books', 'works', and 'authors' oeuvres'.

The concept of the objects of author-title cataloguing so defined, and the denotation of the corresponding term, though still of a highly generalizing nature, stand already on a lower level of abstraction than the first-level ones. From this concept one can already directly infer one of the fundamental and eminently relevant characteristics of all the individual objects which may be subsumed under it, namely, their making their appearance, if not in practice, so at least in principle, always in company: in every single book there are at least two, more often three such objects united. We shall see later on that there exist also other agents involving an amalgamation in a single book of two or even more distinct objects of cataloguing, but here we are concerned only with that particular instance of several such objects being united in one book, which originates in the triplicity of the functions.

This triplicity implies that each book in the holdings of a library has to be recorded in a manner suitable for serving three totally different purposes. If one could achieve this by means of a single device serving the three functions simultaneously, then to distinguish between the three different objects of cataloguing united in a single book would be only a matter of logical nicety. We have, however, as often as not, no such device at our disposal. To answer the different requirements of the three functions, many books have to be recorded in three different ways, by means of three different entries headed by three different combinations of distinguishing marks. The distinction in question becomes, thus, also from a practical point of view, a matter of fundamental importance.

16. FORMAL MARKS, PHYSICAL MARKS, UNIFORM MARKS. THE AUTHORSHIP PRINCIPLE. Let us now inspect the distinguishing marks we make use of when recording the objects of these three different kinds fused in a single book.

We have already said that author-title cataloguing is principally based upon the formal side, the formal characteristics, the formal distinguishing marks of the objects to be recorded. We propose to call these characteristics shortly 'formal marks'. This denomination is intended to contrast the devices by which the author-title catalogue is arranged, with the devices applied for the corresponding purpose in subject and classified catalogues. The attribute 'formal' in this term is justified by the fact that these marks are used as signs, identification tags, attached to the individual objects of cataloguing, and have accordingly, in principle at least, once and for all fixed forms; in contradistinction to the symbols and formulae which the classified and subject catalogues are making use of,that are not based on formal characteristics, but directly on the subject-content of the items they refer to, and that, moreover, vary with regard to the same book according to the different systems upon which these catalogues are based.

We have also pointed out that we are concerned only with autonomous objects. So we must consider here only those kinds of formal marks with the help of which objects are rendered autonomous, i. e., which are used for fixing in the catalogue the place of the records – the entries – of these objects. In the following we shall use the term 'formal marks', therefore, with restrictive denotation comprising only these particular kinds of formal marks.

These kinds of formal marks, of 'books' and 'works' alike, were invented and adopted by convention ages ago. From a logical point of view the first place among them is due to the title which is generally a short and pregnant formula, a label, applied to each individual book and work for the purpose of distinguishing it from other books or works, respectively, and of facilitating its quick and easy identification. But cataloguers were accustomed long ago to employ also another datum: their author's name for identifying as well as locating books and works. They extended in the course of time this practice beyond real authors to the personal editors and compilers of certain kinds of books and works with complex contents, and to the so-called corporate authors, too. The present habit of identifying and designating a book or work by its title and, if it is known, by the name of its author, is thus rooted in a venerable age-old tradition (the extension of this habit for including the corporate author in a considerably less old and venerable tradition). The role played by the title and the author's name of books and works is very similar to that of the modern Western names, the counterpart of the forename is the title, that of the surname is the name of the author. In the course of time the similarity had become enhanced by both having been submitted to an inversion. Already centuries ago bibliographical practice came, for obvious practical reasons, to reverse the genuine order of forename and surname; while in the antiquity the same had been done with the title and the author's name. The latter step had been taken for considerations of the third catalogue-function as well as for expediency in the fields of the other functions, the author's name being as to conciseness and pregnancy generally superior to the title, and in consequence more likely to be remembered. By this reversal of the logical order, the author's name had been allotted the first role, and the title relegated to a secondary one: to the role of determining the place of each author entry within the batch of entries relating to a particular author, and of serving in default of an author's name as a stop-gap.

I should like to point out, parenthetically, that by emphasizing the fundamental importance of the title, and even more so by assigning to it the logical priority, my conception is directly opposed to a central concept of a school of modern cataloguing theory: the authorship principle. In my opinion, the overemphasizing of the importance of authorship which theorists have gradually worked themselves into, and which culminated in the absurd position taken by Miss J. Pettee (cf. Note 54), was not a happy move, theoretically and practically either. It is a pity they had forgotten Cutter's thesis according to which there are two great principles to which almost all cataloguing rules may be reduced – and not one only! The disproportionate stress they had come to lay on the 'authorship principle' implies a corresponding slighting of the title, and has landed them on a false track. There cannot be, after all, any doubt about the title being equal in importance with the author's name, as also Cutter's thesis implies. True, codes do usually start with determining the field of application of author entries and allot to title entries simply the residual area. They are right in doing so, but only for reasons of expediency, and by no means because author entries would be in principle superior to title ones. In direct contradiction to the authorship principle, I am, as I have already hinted at, even of the opinion that so far as logic is concerned, it is the title to which the priority must be assigned. A weighty reason for this is that virtually all books and works are provided with titles, while a great deal of them lack an author. My main argument, however, is that the author's name is a formal mark specific only to his whole 'oeuvre', but not to any particular one among the works that 'oeuvre' is composed of. Not the name Thackeray, but only the title *Vanity Fair* determines the individual work.

The mistake involved in the 'authorship principle' is clearly manifested in its practical and theoretical effects. Practically, it has led to unwholesome efforts arbitrarily to narrow the area of the employment of title entries; it is responsible for the introduction of corporate main entries and for their unsound pullulation; and also for the reluctance to part with editor and compiler main entries. The main harm it theoretically did was to divert attention from the distinguishing marks of books and works to the facts these marks refer to. The first task of the cataloguer is to take account, very carefully, of all the distinguishing marks attached to the item in hand and to ponder on utilising them as efficiently as possible. The facts behind the marks he is only secondarily concerned with. The same applies, of course, to the code-maker, too. Now, under the influence of the authorship principle, code-makers had come to pay exaggerated attention to the facts, the circumstances of the production of the objects of cataloguing. Among others, they had made the intellectual responsibility for the production of these objects – a concept not only extremely obscure, but also of little relevance to cataloguing – one of the most prominent hinges on which their whole system turns. They have overlooked that the 'cataloguing conditions' they must always first get a good grasp of when entering a book, have to be conceived as the *multifarious combinations of the formal marks themselves attached to the items in hand*, and not as the combinations of the facts behind these marks. The distinction is, of course, principally a conceptual, a logical, but nevertheless an important one – had it not been overlooked, the mistakes just pointed out could all have been avoided.

The title and the author's name are far the most important among the formal marks relevant to our subject, but not the only ones. In the denotation of this term we shall include, along with them, also the names of persons who have played certain auxiliary parts in the production of either a book or a work, and the names of bodies that have had something to do with it, as well as subtitles and eventual variants of the title since all of these may help rendering an object of cataloguing autonomous. Later on, when we come to surveying the objects of cataloguing more closely, we shall reach an even more detailed specification of the marks which cataloguers are accustomed to utilize in rendering a book or work retrievable. The same survey will also bring forth a clarification of the denotation which must be assigned to the word 'author' in this context.

The formal marks with which we are concerned fall into two distinct classes. One of these is composed of the formal marks actually borne by books. These are displayed by material objects visibly, physically; and we may call them, therefore, 'physical marks'. Ever since the beginning of a consistent use of title-pages, which lagged only a few decades behind the invention of bookprinting, the physical marks have been the most conspicuous and most strongly emphasized distinguishing marks of books. In spite of this, it took a considerable time until cataloguers came to utilize them as the principal instruments for the designation and identification of books, as well as for fixing in the catalogue the place of the information concerning a particular book. Today, however, they must, without doubt, be regarded as the principal means of performing the first function of author-title cataloguing; one cannot consider this function adequately performed if it is not possible to trace in the catalogue every book of the holdings under its own physical marks. The discussion of the last two propositions had better be postponed until we come to the topic of the dominantly physical nature of most objects of the first function.[17]

The denotation of the term 'physical marks' is not restricted here to exclude marks not displayed by the title-page. Many items a cataloguer must deal with have no title-page, and there are others the title-pages of which do not display all or any of the data our term 'physical marks' embraces. This occurs most frequently with volumes in a series, the title-pages of which very often fail to display the formal marks either of the series or of the work contained in the volume, but it occurs also in a great variety of other situations. In such a case we have to regard that form of a datum which appears anywhere in the document, as its physical mark. If a datum happens to appear in the same document in several different forms, then it is the most strongly accentuated of these forms (e. g., the one appearing in the preliminaries or in the colophon, or else the most often used one) to which we must assign the role of physical mark.

To a catalogue-user seeking not a particular book but a particular work or author's 'oeuvre', the recording of books under their physical marks is, however, not satisfactory at all. The majority of such users want to be informed about *all* the available books containing a particular work, or pertaining to a particular author's 'oeuvre'. (The fact that today we must regard this demand as an absolutely rightful one will be discussed at length later.) The different editions of a particular work and the different books of a particular author bear, however, very often diverse physical marks, so that their entries, if they were headed by these, would be scattered over the whole catalogue. So the needs of the catalogue-users under

[17] See §§ 30–31.

consideration would not be provided for adequately if the catalogue contained only entries under the physical marks of books. Therefore, when a cataloguer wants to record an item, not in its capacity as a detached, independent document, but as an edition of a particular work or a constituent part of a particular author's 'oeuvre', he has to resort to registering it, not only under its physical marks, but also under another set of formal marks, which he applies uniformly in recording also all other documents which contain the same work or a work of the same author, respectively. To distinguish them from the physical marks, we shall call the formal marks belonging to this second class – in harmony with the terminology of the *Paris Statement of Principles* – 'uniform marks'.

The uniform marks are, then, formal marks assigned by cataloguing and bibliographical practice to works and authors' 'oeuvres' for the sake of the second and third functions, respectively. If one considers the matter from the point of view of a particular physical item, of a given book or other document, one cannot deny that the use of uniform marks *is* in a certain sense arbitrary. The disadvantageous effects this is liable to have in regard to the users acquainted only with the physical marks of a document, can be, and in fact usually are, mitigated by choosing for uniform marks the best known and predominantly used formal marks attached to a work or author's 'oeuvre'. These are, as a rule, identical with the physical marks borne by the majority of the editions of a work, or with the name borne by a majority of the books of an author; but there are also exceptions.

The two classes of formal marks agree in their meaning, i. e. they both comprise the titles of the same works and the same authors' and collaborators' names. They differ only in the respective actual forms of many of these titles and names.

17. Characteristics of the physical and uniform marks. The formal marks belonging to both classes are for the cataloguer as well as for the user of the catalogue simply identification tags, that is, signs revealing the identity and facilitating the identification of individual books, works, and authors' 'oeuvres'. Signs must, however, be apprehensible by the senses, in our case they must be visible. Uniform marks, although they become so in catalogues and literary references, are not attached in visible form to the objects which they are to stand for; and it is not even possible that they should be, seeing that these objects (the works and the authors' 'oeuvres') are not of a physical nature. (This is again a matter for detailed discussion, which, however, we must postpone.)[18] The visibility of uniform marks is, thus, lop-sided, and so is also their sign-function itself. As distinct from them, physical marks are visible characteristics of physical objects, identification tags materially attached to the objects which they stand for; in the sign-function and the formal nature of physical marks there is no lop-sidedness at all. In consequence, physical marks are always firmly established, there cannot be any doubt about the physical marks a particular book actually bears. Contrarily, concerning the uniform marks of a particular work or author's 'oeuvre' – since these are not rooted in material reality, merely in convention – opinions often differ and doubts easily arise.

It cannot even be denied that uniform marks are in a certain respect akin to the formulae or symbols employed in organizing the diverse (classified or alphabeti-

[18] See § 30.

cal) kinds of subject catalogues; both being no visible characteristics of the documents catalogued, but derived from, and directly referring to, their intellectual sides. The function of uniform marks consists in bringing together in the catalogue items which are related to each other not formally, not by their visible characteristics, but intellectually, by their contents, or else by the common mental origin of these contents which is also an intellectual and no formal link. As to their function, uniform marks occupy, thus, an intermediate position between the physical marks and the symbols and formulae placed at the head of the entries of classified catalogues and of subject entries. This position corresponds to the likewise intermediate position of the objects of the second and third functions, which will be pointed out in a later stage.[19]

The differences between the two classes of formal marks may be summed up in the statement that uniform marks do not meet the criteria of formal marks in the same degree as physical marks. In consideration of this, one might, perhaps, be tempted to regard our term 'formal marks', on the grounds that it presents the physical and uniform marks as belonging to the same genus, as logically untenable. Such an objection, however, would be unfounded. The use of the term 'formal marks' with the comprehensive meaning we have given it, is warranted on the grounds that between physical and uniform marks there exists, in spite of an essential difference, a fundamental relationship. This consists in their constituting together that genus of devices upon which the organization of the author-title catalogue is based, and in their being, in consequence of their formal character, equally antithetical to the formulae and symbols applied for the corresponding purpose in classified and subject catalogues.

It would also be an error to raise against the latter argument the objection that these formulae and symbols have a no less formal character than the formal marks. By the attribute 'formal' we, of course, do not refer to the formal nature, to the sign-character, of these cataloguing devices themselves – which is, of course, in both cases equally undeniable – but to the nature of the characteristics of the objects of cataloguing from which those devices are derived. In this respect the symbols and formulae of classified and subject cataloguing cannot be called 'formal' any more. They are derived from, and directly refer to, the subject-content, the intellectual constituent of documents; they have to characterize the latter with the greatest possible accuracy. Whereas in author-title cataloguing both classes of formal marks are made use of absolutely irrespective of their aptitude for conveying an idea about the contents, about the intellectual side of the objects they stand for. It is from this side, that the symbols and formulae of classified and subject catalogues present themselves as antithetical to both classes of *our* formal marks, and this antithesis justifies the introduction of the term 'formal marks' as a comprehensive and distinctive denomination for both of these, and only these, classes of catalogue-organizing devices.

Before abandoning this topic, we deem it necessary to emphasize once more that from now on we shall use the terms 'formal marks', 'physical marks', 'uniform marks' only in the sense indicated on p. 36, in spite of the fact that, with the exception of the 'uniform marks', it would be, obviously, possible to give these terms considerably more comprehensive denotations. We have, however, no use for

[19] Cf. p. 58.

such, because the only formal marks which are really relevant to fixing the place of an entry in the catalogue, in other words, to rendering the object of this entry autonomous, and, in consequence, of any interest to us, are the authors', bodies' and collaborators' names and the titles.[20]

18. ELEMENTAL AND COMPOSITE OBJECTS. We may now continue our analysis of the situation created by the triplicity of the catalogue-functions and by the fact that owing to it there are mostly three different objects of cataloguing united in a single book. The next thing we have to point out is that while the first function is atomistic in character, conveying merely isolated bits of information about single, detached items, the second and third functions consist in relating these isolated informational units to each other on the grounds of immaterial, intellectual characteristics of their respective objects, and in arranging, thereby, the whole bulk of the information imparted by the catalogue into a pattern of organically, intellectually coherent groups. The two latter functions create, thus, new and composite units of information, by collocating a number of single informational units every one of which refers, separately considered, to a detached, simple object of the first function. Every such composite unit of information refers to a new object of cataloguing: a 'composite object' of the second or the third function. To contrast them with these composite objects, I am going to call the simple, detached objects of the first function 'elemental objects'.

We shall see later that all the objects of the first function are elemental ones, and that, conversely, elemental objects occur only in the domain of the first function in a broader sense of the word. From now on we shall apply to all objects the communication of information about which belongs to the scope of the first function, the term 'elemental object'. The explanation for our distinction between the scope of the first function in a proper and in a broader sense, and also for our preference of the term 'elemental object' to the term 'object of the first function', we shall give at a later stage.[21]

19. TRIPLICITY OF FUNCTIONS – MULTIPLICATION OF OBJECTS. The proposition in the preceding paragraphs that by the triplicity of the functions a single book is turned into three different objects of cataloguing, is only approximately true and must be qualified in several respects.

First of all, it is obvious that a single book does not constitute three, but only one separate object *complete* in itself: an elemental object. But along with this it

[20] As mentioned before, we are calling an object autonomous if the place of the catalogue entry relating to it is determined by its own formal – either physical or uniform – marks. Now, when several entries in the catalogue are headed by the same name and/or title, the latter are liable to fail to determine the exact place of these entries, which cannot be fixed without taking recourse to further formal marks (e. g., the language of a translation, the year of publication, etc.) of the objects to which the entries refer. This means that along with the authors' and collaborators' names and the titles there are also other marks which may participate in rendering a particular object autonomous. The role of these marks as tertiary and quaternary determinants of the exact place of an entry is, however, a decidedly inferior one, and therefore it is usually ignored by cataloguing theory and codes; it is only the filing regulations which deal with these marks in this capacity of theirs. This is the reason for our attaching to the terms 'formal marks', 'physical marks', 'uniform marks' only the restricted denotations indicated in the text above.

[21] See § 45.

constitutes also one of the components of at least one, mostly of two, but sometimes even of several composite objects of one or of both the other two functions. Secondly, and even more obviously, the third function terminates where anonymous works are concerned; here the absence of the third function involves a corresponding reduction in the multiplication of the objects of cataloguing. We have to mention, finally, that we use the word 'book' here only provisionally, instead of the terms 'primary', 'secondary' and 'tertiary elemental object', by which it will be displaced when we shall have finished delimiting the exact denotations of the latter terms.

With these qualifications, the statement about the multiplication of the objects of cataloguing brought about by the triplicity of the catalogue-functions is generally valid. It also remains valid even if the same physical marks are displayed by each edition of a work, or if the same name of an author figures on all his books, and if such consistently applied physical marks are also unanimously adopted by bibliographical usage. In such instances, the composite unit of information relating to a particular work or author's 'oeuvre' is produced in the catalogue automatically, by the actual uniformity of the physical marks displayed by all the components of the composite object of this unit of information; it comes into existence without any special effort on the part of the cataloguer.[22] We must not allow ourselves, however, by this circumstance to be misled into thinking that in such cases the second and third functions are not performed at all. Even though the triplicity/duplicity of objects is not manifested in the catalogue materially, by an appearance of three/two diverse entries, it is existent notwithstanding: the catalogue does not fail by any means in these cases either to communicate, along with the information about the detached elemental objects, also an organized one about the works and the authors' 'oeuvres' which these elemental objects unite to form.

There arises, however, in this connection a question which it is less easy to answer: whether the triplicity of the functions and objects does or does not persist in the case of a work which has appeared only in a unique edition, and in the case of an author who has published only a single book. From the proposition that the two last functions consist in collecting several units of information, one seems to be obliged to infer that to such a case these two functions do not apply at all, and that here one has to deal only with one object of the first function.

There is, however, also another side to this matter – even irrespective of the formal argument that the catalogue in fact *does* communicate information, in such cases too, about *all* the editions of the work and *all* the books of the author. It happens very often that a library possesses only one of several existing editions of a particular work, or of several books published by a particular author. If we adopted the position that it is an indispensable feature of every individual object of the second and third functions to be built up out of several elemental objects, we should have to conclude that in the library in question the solitary books mentioned do not become objects of these two functions, and that, therefore, many books which in well-equipped libraries become triple/double objects of cataloguing, remain in less well-equipped libraries simple objects only.

[22] If with a work the author of which is known the actual uniformity here mentioned is confined to either the title or the author's name only, the recording of the three different objects requires, naturally, two different entries.

That this is not a sound inference becomes quite evident if we consider the widespread actual usage of cataloguing a book with physical marks different from the uniform marks usually attached to the work it contains, or to the author's 'oeuvre' it belongs to. Two libraries which both accept the three functions are bound to render such a book traceable in just the same way, under both kinds of formal marks, even if one of them possesses several editions of the work contained in it or several books by its author, at least one of which bears non-uniform physical marks, while the other is in possession of only one edition of the work in question, or of one book by this author, bearing non-uniform marks. The latter library will not omit, either, to make that book traceable under both sets of formal marks. That is, it will not fail to discharge the second and the third functions in this particular case either, and will behave as if it, too, had to collect several components of a composite object displaying different marks. It will take this course partly for the sake of a consistent adherence to international usage, but partly also for its own sake, because it has to reckon with the possibility of acquiring in the future another edition of the same work, or another book by the same author, which may display physical marks different from those borne by the solitary item it already contains. It is an unquestionable fact of present prevailing cataloguing practice that all works which have appeared in more than one edition, and all the 'oeuvres' of authors who have published more than one book, to which at least two different sets of formal marks are attached, are treated, without regard to the actual holdings of the library, as triple/double objects of cataloguing, by every cataloguer not rejecting the second and third functions.

From this one must, of course, infer that a book containing a work which appeared only in this one edition, and a book by an author who has published no other books, cannot make any exception to the rule that every book is an object of three/two diverse functions and unites three/two diverse objects of author-title cataloguing. The discovery of another edition or the publication of a new edition is always possible, which establishes an analogy between the absolutely unique edition and that which is unique only within the holdings of a particular library. The same applies also to the uniqueness of a book by a particular author. The similarity is increased further by the not entirely impossible contingency that even a unique edition of a work, or a unique book of an author, might be better known by some conventional marks than by its own physical ones, and may have to be recorded, therefore, under two diverse sets of formal marks.

I think this similarity far-reaching enough to settle conclusively the question under consideration. At any rate, it yields a weighty argument in support of the proposition – with only the clearly defined two modifications specified at the beginning of this paragraph – that the triplicity of the functions *always* entails a triplicity/duplicity of the objects of cataloguing. It is important to stress this point as this proposition constitutes one of the cardinal points of our classification of the objects of author-title cataloguing, and so the simplicity and perspicuity of the latter depend in a marked degree on its unimpaired validity.

It is necessary to emphasize that along with the statement that every elemental object of cataloguing becomes automatically also an object, or more precisely, a component of an object of the second and mostly also of the third function, the proposition under analysis contains also the statement that it is impossible for any item to be catalogued as a component part of an object of the two last functions without being recorded at the same time also in its capacity as an elemental object. The necessity of emphasizing this point does not become apparent within the limits set by the current formula for the first function, i. e., until it is supposed that there do not exist any elemental objects of standard cataloguing other than the books in the broadest possible sense of the word. Later on we shall see, however, that this conception is in need of modification, in consequence of the fact that actually there are whole classes of elemental objects which are not included even in the most comprehensive of the acceptable denotations of the word 'book'. Among these there happen to be several classes with regard to which it might seem plausible to assume that they are recorded in the catalogue only on behalf of the second and third functions, i. e., only in their capacity as components of an object of one or of both of these functions, without becoming at the same time also separate objects in their own right, i. e., elemental objects, objects of the first function. Such an assumption would be certainly erroneous. The truth of this will become apparent only when these non-book-objects of standard cataloguing have been subjected to closer inspection.[23] Anticipating this verification, we shall henceforward refer to the proposition that in a particular catalogue no autonomously recorded item can be found which is not also one of its elemental objects, as if it has already been verified.

20. A THIRD DENOTATION OF THE TERM 'OBJECT OF CATALOGUING'. We have come to the end of the discussion of the second denotation of the term 'object of author-title cataloguing'. The determination of a third denotation, a survey, specification and classification of those particular kinds of objects this third denotation comprises, will form the subject of the best part of this inquiry. We shall have to take stock of all the sub-classes the three main classes of objects (books, works, authors' 'oeuvres') in turn are composed of, fixing at the same time with every possible accuracy the limits of each of these classes, i. e., the limits of the second denotation of the term 'object'. The solution of this task will yield thus also, as far as it is possible, an accurate delimitation of the functions of author-title cataloguing, i. e., the attainment of our final objective.

We have already mentioned that with regard to the objects of author-title cataloguing we cannot rely on such familiar ideas and widely accepted formulae as in the case of the functions. So we shall have to try, first of all, to give an account of the actual state of things, of the objects which present-day practice is in the habit of recording; but occasionally we shall also attach some critical remarks. In our account of the actual conditions we shall employ the method of assuming – provided that there is no specific reason for not doing so – first, that the codes of a few great countries reflect the practice actually followed in these countries regarding the range of objects usually recorded in author-title catalogues; and second, that the practice indicated by the majority of these codes is representative of the actual world-wide state of cataloguing. To the standpoints of minorities we shall refer only in exceptionally warranted cases. This method will be employed also in examining the question how far the current formula of the functions corresponds to present-day actual cataloguing practice. A broader basis of observation would involve us inevitably in irrelevant detail.

[23] See Chapter V.

CHAPTER III

A BRIEF HISTORICAL OUTLINE OF THE DEVELOPMENT OF THE FUNCTIONS OF THE AUTHOR-TITLE CATALOGUE

21. Earliest stage: Neglect of first, approximation to second, fulfilment of third function. Before proceeding to analyse the three functions of today's author-title catalogue in turn, we had better throw a cursory glance at the European-American tradition, at the development which has led to these functions and to the modern views on them. One of the factors which play, at any time, an outstanding role in fashioning actual catalogue-functions is, certainly, the past, so it is impossible correctly to explain and to assess present-day functions without taking into consideration at least the chief traits of the development which has led up to them.

For the roots of the functions of our present-day catalogues we must look in the cataloguing practices of the late 15th century. It was at that time that the problems modern cataloguing is still struggling with, first emerged, owing to the appearance of printed books and to the sudden growth of library holdings brought about by the quick spread of the art of printing. It should not be forgotten, however, that everything we shall say about 15th and 16th century cataloguing practice has to be taken *cum grano salis*, as the cataloguers of these early times did not display much consistency in their application of the principles and rules underlying their practice.

The greater part of the catalogues of this epoch may be characterized as having been records of works rather than of books. Their entries were generally composed of not more than the conventional, the best-known form of the author's name and of the title of the work contained in the book referred to; the title having been usually cited, even in case of books in the vernacular, in its Latin form. These entries neglected, thus, to indicate the form of both titles and authors' names actually figuring in the books themselves and differing from the conventional one. They did not contain any further data, either, which could have helped to identify the books they related to. The cataloguer of the 15th century even deemed it unnecessary to mention whether the item he recorded was a manuscript or a printed book. Cataloguing several different books, he made not rarely the very same entry several times – sometimes even in an uninterrupted sequence – without adding any distinctive data, e. g., the place or date of publication. That is, he did not care to apply the devices of today's descriptive cataloguing (in the broader sense of the word) for distinguishing different books containing the same work; in other words, according to our present-day notions he did not care to record 'books'. All types of book-lists of this epoch which were to some extent related to the modern author-title catalogue followed this usage: the alphabetical lists and indexes, as well as the more frequent type of catalogue which was divided into a few great subject groups and subarranged within these groups alphabetically. An attempt to distinguish by descriptive means an individual book from all other books was made most frequently in inventories and shelf lists. But the means employed in them for this purpose (remarks concerning the size, the binding, the hand-writing, or else a quotation of the incipit, the explicit, or the first words of the second or the penultimate leaf) were fashioned almost without exception so as to suit codices, and were in consequence either entirely unfit to characterize a particular printed edition of a given work and to facilitate its identification, or they performed these tasks only very inefficiently. The distinction and identification of books by means of individual location marks, in fact applied in many medieval libraries, cannot be regarded by any means as a device fit to fulfil the first function as it is understood today. In our days we do not consider this function as adequately performed unless the catalogue gives a definite answer to the question whether a particular book is contained in the library or not, an answer which must be unequivocal in itself, without any recurrence to the book described, and besides, directly or indirectly traceable under the physical marks of this book.[24]

All this does not mean, naturally, that the 15th century librarians deliberately resolved to catalogue works only, and not books. And it would be an even greater mistake to assume that they recognized the duty of securing easy access to all the available editions of a work, i. e., of fulfilling what we are calling now the second function. Very probably, it never occurred to them to collocate deliberately the entries of all the manuscripts and printed books containing the same work; only a few anonymous works, first of all the *Bible* and some liturgical works, were exceptions. Notwithstanding that, the cataloguing of the works entered under authors' names in fact approximated a fulfilment of this task, as the tolerably consistent use of conventional, i. e., uniform, authors' names yielded as a by-product automatically also a rough relating and displaying together of all their available editions.

So, rather paradoxically, in spite of the fact that their practice may be characterized as a recording of works rather than of books, they fell also short of recording works in the manner which in our own days is considered indispensable to the fulfilment of the second function. Up to the beginning of the 16th century the forerunners of the author and title catalogue approximated the performance of only one of the three functions of our present-day catalogue: that of the third function, of the function to communicate information, within the limits of the holdings of the library, about authors' 'oeuvres'. Even this was carried out, however, inconsistently, and was not based on conscious deliberation. Most probably it did not even occur to the average 15th century librarian that it would be possible to treat authors' names in any other manner.

So cataloguing was rather an instinctive affair at that time. Very little speculation was spent on analysing either its objectives or its appropriate means. This must not astonish us at all as the tiny holdings of libraries and the modest volume of universal literature did not yet require any differentiated informational tools

[24] All these features of the first function of present-day author-title catalogues will be discussed in the proper contexts at later stages of this inquiry.

and cataloguing methods. The choice of headings and entry words was a matter of incomparably less importance than it is today, and so the lack of entries with headings strictly corresponding to the physical marks of the books did not entail any harm worth mentioning. There simply did not yet exist any inducement to fulfil the requirements which the first function, in the modern sense, includes or even to recognize the utility of doing so. A subarrangement by works of the entries under a particular author's name was not necessary as yet; the medieval librarian would have thought the idea to regard this as a separate task, different from the task of conveying information about the 'oeuvres' of authors, on rather good grounds, unpardonably pedantic. There was surely no practical inducement for his distinguishing the two tasks from each other, for discovering their logical separateness.

22. Second stage: The first function gains ground. Th. Hyde. In the 16th century the practice of appending to the entries of printed books a statement of the place and year of publication emerged. Some authors maintain that the objective of this innovation was simply to indicate that the entry referred not to a manuscript, but to a printed book. Be this as it may – relevant to us is only the fact that this innovation was actually the first step taken towards the adoption of what we are calling today the first function, and also towards a replacement of the work by the book in the role of the primary object of author-title cataloguing. This step did not involve any discarding or eclipsing of the functions which had already been performed before. In the quickly-spreading author and title catalogues authors continued to be entered under the conventional form of their name, that is to say, the third function continued to be fulfilled almost adequately, the second one approximately. Throughout the 16th century the prevalence of conventional formal marks does not seem to have been mitigated even by a secondary recording of physical marks. In the 3rd edition of his bibliography in 1583, Conrad Gesner, though making a rather liberal use of name-references, did not show any sign yet of recognizing the importance of a consistent recording of the physical marks of books, but it must be admitted, that this did not perceptibly hamper the use of his work.

In the 17th century, however, along with the quickening growth of the volume of literary production, especially in the vernacular, the difficulties of tracing a particular book arising from the neglect of physical marks, must have already become apparent, and this led to a gradual spread of the practice of registering, along with the conventional marks of 'works', also some physical marks of 'books'. As early as 1674 a librarian already contrived to formulate very articulate rules providing for a consistent recording of such variant forms of authors' names different from the conventional name, which did actually figure in the books catalogued.

This librarian was Thomas Hyde, who in the preface of the 1674 edition of the catalogue of the *Bodleian Library* stated the rules according to which the authors known under more than one name or form of name had been handled in this catalogue.[25] Though the wording of these rules is not always unequivocal, yet, if one considers the catalogue's practice too, it becomes quite evident that regarding

[25] Hyde, Th.: *Catalogus Impressorum Librorum Bibliothecae Bodleianae in Academia Oxoniensi*. Oxoniae, 1674, Praefatio.

authors whose name was used in several different forms, Hyde's objectives were twofold: to assemble, in harmony with traditional practice, all entries of books by the same person under a particular form of his name, and to register, besides, all other forms of that name occurring in the books catalogued, by referring from them to the preferred *variant*. His intentions concerning authors with several different *names* are less clear. This does not diminish, however, the merit of his having been, so far as I know, the first explicitly to formulate definite rules providing for the use in main entries of uniform authors' names, as well as for the obligatory recording of the variants of names displayed by the books catalogued. These rules contain a clear-cut statement of policy, at least so far as the books entered under their author's name are concerned, that includes a systematical and consistent performance of two distinct functions – the first and the third – of the modern catalogue. Still, it is problematical whether or not it had occurred to Hyde that his catalogue actually performed two distinctly separate functions.

Similarly unclear is his attitude towards the second function, the function of providing pooled information about all the editions of a work contained in the library. By assembling under a uniform name all the entries of books by a given author, so far as authors' works were concerned, he actually did fulfil, although somewhat crudely, for all practical purposes this function, too. But whether he recognized that this was a function distinct from communicating information about authors' 'oeuvres', we may only conjecture, in default of an explicit statement by him. One or the other among his rules might be interpreted as implying the aim to collect and relate all the editions of anonymous works, too, but they might also be interpreted contrarily, perhaps with just as good reason. It is probable that Hyde did not recognize yet the second function as a separate task. In his days the holdings of the Bodleian were not large enough to render a strict subarrangement, according to his works, of the entries made under a particular author's name a matter of practical importance, nor to give rise to the thought that what we call nowadays the second and third functions, was not a single function only. As regards a few frequently published anonymous works, Hyde made use of the traditional device of entering them under their conventional titles; but it is more likely than not that he accepted this inveterate solution without much speculation, and failed to discover that the collocation of the diverse editions of a particular work forms a separate task requiring a general and uniform solution.

23. Third stage: Gradual recognition of the second function. The practice of referring from the author's name figuring on a particular book to the uniform name under which all the entries for this author are collocated, came to be generally adopted during the two centuries following the publication of Hyde's rules. At the same time, in the rather protracted development of the rules for cataloguing anonymous works there was becoming discernible – very slowly, but with a gradually increasing definiteness – a tendency to pool the information about the different editions of a work, that is to say, a tendency to accept and recognize the second function as it is understood today. A very important step in this direction was taken by Jewett in his rule XXIII, according to which "The translations of works of unknown authorship are to be entered, like other anonymous works, under the first word of the original title, whether the original be or be not in

the library".[26] The first author, however, who fully recognized the second function was Cutter, and even he was approaching that standpoint only step by step.

Let us point out in passing that the extraordinary protractedness of the process finally leading to the acceptance of the second function is a striking illustration of the fact that our forerunners – resembling in this respect the majority of our contemporaries – did not worry much about examining their own practices theoretically. A striking instance of this negligence was their prolonged reluctance to extend to *all* anonymous works the method applied already since the Middle Ages to the *Bible* and a few other frequently published anonymous works, i. e., to find their way from a sporadic to a general and consistent pursuit of even an outstanding objective.

The position held in regard to the functions of author-title cataloguing by the international practice of our days had been reached by the countries most advanced in librarianship at the turn of the century, when the German and the Anglo-American codes had been first published. Along with the obligation of conveying adequate information about each separate book, both codes also explicitly adopt the principle of a compulsory application, not only of uniform authors' names, but also of uniform titles. This means that they cogently prescribe the performance of all the three functions.

Our present-day views on the functions of author-title cataloguing came to be firmly established, thus, only at a rather recent date, although the beginnings of the development leading up to them lay several centuries back. The first stage in the first phase of this long-draw-out process was marked out by a tacit application, the second stage by the explicit formulation, of cataloguing rules only *implying* the performance of these functions. As usual, practice had produced, here too, appropriate solutions and rather satisfactory usages, long before theory managed to become aware of the task to be solved.

24. FOURTH STAGE: EXPLICIT STATEMENT OF THE FUNCTIONS. C. A. CUTTER. It is from the time when librarians began to show their awareness of the theoretical aspect of the problem, that we may reckon the second phase in the history of the functions of author-title cataloguing. In this phase the decisive step was taken by Cutter. He was, to my knowledge, the first to care explicitly to state the objectives, the functions – or as he termed, the 'objects' – of the catalogue for which he had set out to compile a code. This having been the dictionary catalogue, only three among his propositions concerning its functions apply to author-title cataloguing. Two out of the three – those concerned with the information about books and authors' 'oeuvres' – are endorsed by an overwhelming majority of librarians even in these days. The third, according to which the catalogue should assist its users "in the choice of a book as to its edition", though it refers to a point which forms a part of the second function, cannot be considered a quite unequivocal statement of the adoption of this function as it is understood today.[27]

[26] JEWETT, CH. C.: Op.cit., p. 53 (See Note 2).

[27] CUTTER, C. A.: *Rules for a Dictionary Catalog.* 4th ed. Washington, 1904. Republished by the Library Association, London, 1953. – Cf. p. 12, point 3/G in the specification of the 'objects' of the catalogue.

Cataloguers have by now firmly established and generally accepted definitions of all three functions. This does not mean, however, that librarians all over the world in fact agree to accept without reserve either these definitions, or the functions so defined. As we all know, there are still many among them who do not acknowledge the obligation to collocate the information on the different editions of a work, and some reject even the collocation of the information about all the productions of an author.

Another point which I am well aware of is that the demand for the performance of one or the other of the three functions may decline, or even completely vanish in the future, in consequence of improvements which will have been achieved in our bibliographical equipment, or in subject cataloguing, or else, perhaps, in consequence of the invention of new tools only dreamt of, or not even dreamt of, in these days. Nevertheless, there remains the fact that the recent trends in practice and the latest contributions to theory are nearly unanimous in recognizing these three functions. It is true that this unanimity vanishes as soon as the question of practical precedence, i. e., the question as to the respective roles to be assigned to the three forms of entry (main entry, added entry, reference) in performing the functions, is raised; yet, as compared with the actual recognition of the functions, this is a matter of minor importance only. Besides, a trend towards unanimity is perhaps even in this respect already faintly discernible.

25. THE FORMAL CHARACTER OF THE CATALOGUE BECAME CONTINUALLY MORE AND MORE MARKED. From this sketchy historical outline we may draw three general conclusions. The first is that the aim of performing systematically and consistently the three now already widely recognized functions had emerged only at a rather late stage of the history of cataloguing, and that the attempts to define these functions and clearly to distinguish between them were of an even later date. The second is, that the chief cause of this protraction was simply the fact that librarians were very slow in realizing the plurality of the tasks which they actually kept striving by means of the author-title catalogue to solve.

The third conclusion relates to the effect which the secular changes undergone by its functions have produced in the development of the general character of the catalogue. This effect consisted in rendering the catalogue more and more manifestly formal in its character. There have been two particular factors which played prominent parts in bringing about this change.

The first of these factors was a gradually increasing stress laid upon the physical marks of books, a steady enhancement of the role assigned to these marks in cataloguing. Its effect was intensified by the fact already mentioned, that the

←

We may mention here the interesting fact that as early as in the 17th century one found in library literature already a sentence which is surprisingly in harmony with the present-day definition of the second function. This sentence is to be found in FRÉDÉRIC ROSTGAARD's study *'Projet d'une nouvelle méthode pour dresser le catalogue d'une bibliothèque selon les matières'* (2nd ed. Paris, 1698), and runs as follows: ". . . on disposera dans le Catalogue [les livres de la Bibliothèque] d'une manière, que . . . toutes les Éditions d'un même ouvrage s'y trouvent toûjours ensemble." Rostgaard's concern had been, however, not with the author-title, but with the subject catalogue, and so his maxim has no bearing upon our topic. The same applies also to his proposals concerning the means he has intended to use to establish the so circumscribed order in his catalogue.

physical marks fulfil the criteria of formal marks in a higher degree than the uniform ones.[28]

The second of the factors in question was a gradual elimination from the author-title catalogue of headings and entry words derived not from the formal marks but from the subject-content of documents. In earlier centuries such headings and entry words were used rather freely in author-title cataloguing. The adoption of the usage of entering an anonymous work lacking a generally known conventional title under an entry-word which characterized concisely its subject-content, and was either a catchword taken from the actual title, or chosen in the way subject headings are chosen nowadays, was explicitly stated by the English bookseller Andrew Maunsell already in 1595. In the 17–18th centuries this usage became quite widespread.

It is perhaps not superfluous parenthetically to point out that, although by its introduction author-title cataloguing was undeniably trespassing on the domain of subject cataloguing, this usage must by no means be assessed as the adoption of a fourth function of the author-title catalogue, consisting of a deliberate sporadic shouldering of the function of subject cataloguing. The end of our predecessors was merely to improve the fulfilment of the traditional functions of author-title cataloguing. This is clearly shown by the fact that the usage had neither an underlying principle, nor any recognizable boundaries, both of which any attempt at subject cataloguing could not have dispensed with. It depended entirely on chance, on the judgement and taste of the individual cataloguer.[29]

In the 19th century this usage began steadily to decline. As a result of efforts to draw a clear boundary line between the respective spheres of author-title and subject cataloguing, in our century the elimination from the author-title catalogue of headings akin to subject headings continued. This contributed considerably to the character of this catalogue becoming more pronouncedly formal.

The progress achieved in the latter respect is, however, as yet far from being as satisfactory as the one attained in regard to the consistent recording of the physical marks of books. Most present-day codes – including the new *Anglo-American Cataloguing Rules* too – continue to contain rules providing for 'form headings' and other sorts of headings which are in fact mere subject headings (or sub-headings), and to preserve thus the relics of out-of-date practices.

Notwithstanding this, as a result of the joint influence of the two factors mentioned, the formal character of the author-title catalogue is in our days more marked than ever, but it is still rather far from being perfect. Of the three functions of the catalogue only one, the first, is based on a purely formal foundation, the two others impair its formal character considerably. Nevertheless, the term 'formal catalogue', proposed by H. Trebst 40 years ago,[30] suits the modern author-title catalogue very well indeed; whereas its application to the 16th century predecessors of this catalogue would be rather unwarranted.

[28] See § 17.

[29] VERONA, EVA: A Historical Approach to Corporate Entries. *Libri*. Vol. 7, No. 1, 1956, pp. 1–40.

[30] TREBST, H.: Der heutige Erkenntnisstand in der Formal- und in der Sachkatalogisierung. *Zentralblatt für Bibliothekswesen*. Vol. 51, 1934, pp. 441–442.

CHAPTER IV

THE FIRST FUNCTION
I: THE RECORDING OF 'BOOKS'

26. THE SPECIFYING AND CLASSIFYING OF THE OBJECTS OF THE FIRST FUNCTION IS A CARDINAL TASK OF THIS INQUIRY. As already mentioned, for fulfilling our task proper, the task of analysing the three functions and of determining with all possible precision their scope and boundaries, we must first of all survey systematically their respective objects. The expression 'objects of a function' is used here, for brevity, to designate – in accordance with the above specified third meaning of the term 'objects of cataloguing' – those objects about which information should be conveyed within the range of a function.

The survey and the classification of the objects of the first function – the elemental objects – which we now embark upon, constitute one of the central themes of this inquiry. Since the aggregate of all the components of all the objects of the second or of the third function is, as pointed out above,[31] in any given catalogue dependent on the aggregate of the objects of the first function, the following survey and classification of the latter will yield simultaneously also an answer to one of the cardinal questions which will arise in the course of the scrutiny of both other functions.

Let us start from the 'current formula' of the definition of the functions. According to this, the first function consists in efficiently answering the question whether a particular book is contained in the library or not. This means that the catalogue must be capable of giving quick and dependable information on the availability of each book which *is* contained in the holdings, as well as on the non-availability of any other book existing. The latter requirement is, however, fulfilled automatically, if one can rely on the catalogue answering the former dependably and efficiently.

Our first task consists in clarifying the specific meaning the word 'book' in the current formula actually bears. This meaning must, obviously, comprise all the objects of the first function and only these, which means that it cannot be congruous with the everyday meaning of the word. In a note to its definition of the first function the *Paris Statement of Principles* stipulates that "the word book should be taken to include other library materials having similar characteristics". The task before us may be defined as the clarification of the exact meaning of this rather vague stipulation, i. e., as an investigation into the nature and range of the

[31] See § 19.

things the word 'book' stands for in the definition of the first function given by the *Paris Statement of Principles* as well as in what we have proposed to call the 'current formula'.

Preliminarily we must stress a negative point: that in this connection the denotation of the word 'book' does not include either 'works', or 'copies'. Works are, as the current formula shows, objects of the second and not of the first function. The drawing of the line of division between 'books' and 'works' is, as we shall see later on, a task of some intricacy, the solution of which we may, however, postpone, contenting ourselves with a provisional acceptance of the proposition that things lying on the 'work-side' of this line are not included in the denotation of the word 'book', and, therefore, they are outside the scope of the first function. With copies, on the other hand, this study is not concerned at all. The theorist of cataloguing regards an individual book as a unique phenomenon, in spite of its belonging to an edition which consists of thousands of closely similar copies. He is justified to do so, since the structure of the entry proper of a particular book remains the same entirely irrespective of the number of copies possessed by the library. The devices appended to the entry proper for locating all available copies are no integrant parts of the entry, do not touch upon its structure, and are therefore of no relevance to him.

27. THE MEANING OF THE WORD 'BOOK' RELEVANT TO OUR SUBJECT. In turning, now, to the task of determining *positively* the specific denotation to be assigned to the word 'book' in the first proposition of the current formula, we had better restate first of all what we have already found out about this subject in the first two chapters.

The current formula is fashioned so as to suit standard cataloguing. Consequently the word 'book' in it must be interpreted as a comprehensive term which is intended to include all the objects that belong within the scope of the first function of standard cataloguing. Now, when in the course of specifying our intentions regarding scope and limits of this study we have drawn a line of division between those objects which are to be considered objects of standard cataloguing and those which are not, we have made in § 6, and shall make also in Appendix I attached to this paragraph, some statements which are directly relevant in the present context, too. We have said that all printed documents using written language as a means of expression (including leaflets, pamphlets, and ephemera) if actually treated according to the rules of standard cataloguing, as well as all kinds of serials, *do* belong to the range of the objects of standard cataloguing whereas all printed documents using media of expression other than written language, and also those which for one reason or other are catalogued according to methods either more or less elaborate than the standard ones, *do not* belong to it. This applies without any qualification also to that particular denotation of the word 'book' which we are here trying to define. This is to be extended beyond the everyday denotation of the word so as to include all printed documents which actually *are* catalogued according to the rules of standard cataloguing, even if in common speech they are *not* called books; and reversedly, it is to be narrowed so as to exclude all documents which are *not* catalogued according to standard methods, even if they *are* commonly called books.

It is, however, not only in comprising a different assortment of types of documents that the denotation of the word 'book' in the current formula diverges from the everyday meaning of this word, there is also another and even more significant difference between them. For illuminating it, we must first inspect more closely the everyday meaning of the word 'book'.

In everyday speech the word 'book' designates, clearly, a compound phenomenon, composed on the one hand of a material thing, an industrial product, and on the other hand of an intellectual, immaterial component: its subject-content. Another important feature of its everyday meaning is that an object so denominated must always be separate physically, and may occasionally even consist of several such physically separate objects (multi-volumed books). The denotation of the technical term 'book' which we are in need of and now engaged in specifying, although it diverges in several other respects from the everyday meaning of the word, harmonizes with it in these two respects: it also implies these two features.

At this point, however, the congruence of the two meanings already comes to an end. The feature that one book may consist of several physically separate pieces, is encountered in the case, first, of simple books published in several fascicules (*Lieferungswerke*) or volumes (multi-volumed books), and second, of all sorts of serials. So the phenomenon of a single unit being composed of several physically separate pieces appears in many individual cases repeatedly, on two different levels. For instance, there are several issues, not considered books in either the everyday or the technical sense of the word, which constitute an annual volume of a periodical that is in its turn considered a book in both senses; but in a technical, bibliographical sense the whole set of the annual volumes of this periodical is considered another single unit, another single 'book' in the sense of the *Paris Principles*. This is a considerable extension of the everyday meaning of the word 'book', definitely implied in the way this word is employed in the current formula for the first function. On the first level the physically separate parts constituting one bibliographical or catalogical unit are even by their own publishers mostly not considered separate, self-contained publications. It is only technical or commercial considerations, or reasons originating in the character and purpose of the publication, which have led to their having been published in several physically separate parts successively. In these cases the cataloguer and the reader, though faced with several distinct physical pieces, look upon these as upon a single book. On the second level – and this is the relevant thing in this context – the physically separate parts of a serial which may be called a 'book' in a catalogical sense only, have two opposite aspects: they are on the one hand complete books in themselves, self-contained from the publisher's as well as the cataloguer's and reader's point of view; and at the same time they are mere incomplete parts of the serial which is also to be considered a single complete unit, a single 'book', but only in the sense of the current formula.

The fact that the separate pieces in question constitute merely the parts of a single catalogical unit, is manifested in the case of both classes by unmistakable physical (visible) sings: by physical marks carried by all the parts uniformly, by other physical characteristics, like numbering, etc., and in some cases also by a conspicuous incompleteness of the parts (lack of title-page, continuous pagination).

28. A SINGLE 'BOOK' IN THE EVERYDAY SENSE IS OFTEN A COMPOUND OF SEVERAL 'BOOKS' IN THE CATALOGICAL SENSE. The feature that a single physical unit (or a multi-volumed single book) may comprise two different books – more precisely, one complete book and one of the parts of another 'book' – in the catalogical sense, marks the most significant divergence of the peculiar catalogical denotation of the word 'book' from its everyday meaning, a rather sharp divergence at that. Let us inspect this feature more closely.

We have already mentioned that, in order to perform its first function, the catalogue must record each book under its own physical marks; we shall give the reasons for this presently. The double books we have spoken about in the last paragraph display two different sets of physical marks which cannot be mixed up with each other, since one of them belongs to the whole serial, the comprehensive 'book' composed of several physically and often also intellectually separate parts, the other only to one (or to some) of its parts. To this class of double books belong first of all the single numbers constituting the parts of all series and of some collections, and then those issues of periodicals which bear, apart from the physical marks of the periodical, also physical marks of their own which they do not share with the other issues of the periodical. Seen from the point of view of the first function, a whole series and one of its volumes containing a monograph, or again, a periodical and its special issue offered to a scholar as a memorial volume, are from the cataloguer's point of view quite indisputably two entirely different items; their own respective sets of physical marks render each of them an object of the first function distinctly separate from the other, though they are united in a single physical item. It is indispensable that each should be recorded separately. This obligation also includes the autonomous recording of incomplete serials, even of those which are not intended to be completed. Cataloguers do not hesitate to record the defective or incomplete simple books held by their library; they must not hesitate to record the defective or not completed serials either. It is also possible that a single physical item (or a single multi-volumed book) may unite in itself three (series, subseries with a distinctive title, monograph), or even four different 'books' in the catalogical sense. It is indispensable that each of these should be recorded separately that they all should be made autonomous, i. e., elemental objects of cataloguing.

29. WORKS PUBLISHED IN COLLECTIONS ARE MOSTLY LEFT UNRECORDED BY OUR CATALOGUES. One point which in the foregoing paragraph we only touched upon needs to be examined particularly. The occurrence of that amalgamation of two or even more diverse books in the catalogical sense which we have just spoken about, is confined exclusively to physically separate pieces. Let us illuminate the curiosity of this fact by a few examples.

Take a series of papers published periodically by a learned society, each annual volume of the series consisting of 12 separately issued papers, every one of which – whether or not fit out with a separate title-page – displays also physical marks of its own. Most cataloguers insist on cataloguing each of these papers separately too, according to their own physical marks, even if at the end of a year preliminaries are issued for the whole annual volume which merge the 12 separate monographs into a single new, comprehensive catalogical unit. If, however, the very same 12 monographs have never been issued separately, but were already at their publication merged into an annual volume, most cataloguers would abstain from recording them separately. Or again, generally the cataloguer is contented with recording an anthology with the title *The Ten Best Short Stories of the World* by its comprehensive physical marks only. If, however, the very same anthology is published, though under the very same comprehensive title, but in 10 separate booklets, most cataloguers will think it necessary to record each short story separately too, under its own physical marks.

In both of these examples the subject-content of the publication, as well as the diverse sets of physical marks borne by the comprehensive publication and by its constituent parts, respectively, are exactly identical, and in spite of this the manner in which these publications are usually catalogued varies. So even the identity of subject-contents joined with an identity of the physical marks does not suffice to ensure identical treatment. Publications of the above kind become double objects of cataloguing only if their intellectually separate and self-contained components are published as physically separate pieces. Whether or not such a component obtains the status of a separate cataloguing item, of a separate elemental object, depends, thus, entirely on the material criterion of physical separateness. As a consequence, most catalogues render, in a surprisingly inconsistent manner, one part of the available editions of many a work retrievable, another part not.

The curious feature of author-title cataloguing we have come here across, is, as everybody knows, not confined to works published once in a physically separate form, and at another time as a physically not separate part of a publication with complex contents. A much more frequent form of its manifestation is the treatment of works which have never been published in a physically separate form; such works, though self-contained and independent intellectually, do not attain the rank of an autonomous object at all.

Is this a satisfactory state of things? After all, it is their subject-content for which almost all catalogue-users want to obtain books. It would seem, therefore, right that the information to be conveyed by the catalogue, and also the selection and range of the objects of cataloguing, should be dependent first of all on the content, the intellectual side, and not on a material feature, of these objects. Is it not a downright mistake to allow physical separateness to become a factor in such a measure decisive?

30. THE RESPECTIVE IMPORTANCE OF THE MATERIAL SIDE OF THE OBJECTS OF THE THREE FUNCTIONS. The question closing the foregoing paragraph gives rise to another, more general and comprehensive problem. What is in fact the character of the objects which the author-title catalogue is conveying information about: a material, an intellectual, or a dual one? If the last should prove to be the correct answer, then a further question must also be raised concerning the importance for cataloguing of these two different aspects of the objects recorded, and their comparative influence upon the substance and range of the information to be conveyed by the catalogue.

The answer to the first question is simple enough. The objects of author-title cataloguing – with the exception of one class of minor importance which we shall deal with at length at a later stage –[32] are of a dual, a compound nature; they all have two components, two aspects relevant in the present context: a material and an intellectual, immaterial one. That the thing the word 'book' designates is of such a compound nature, we have already mentioned earlier. In using this word we always presuppose the presence of a material component as well as an intellectual one, because a book without a material body is just as unthinkable as a book without contents; in this sense the very word 'book' already implies duality. The same applies also to the objects of the other two functions. Seen from its material side, an object of the second function is – crudely, provisionally defined – a selection of books interconnected by the circumstance that each of them contains the same work; seen from the intellectual side, that object is simply this very work. The material side of an object of the third function is also constituted by a batch of different books: the books by a particular author the library possesses; its intellectual side by that part of the 'oeuvre', of the intellectual production of this author, which is contained in that batch of books.

The two sides are not of equal consequence for cataloguing, and besides, their comparative significance varies with the functions of the catalogue. So the answer to the second of our two questions is liable to become more complex than the one given to the first has been. It is, therefore, advisable to give our representation of the differences between these two sides a greater plasticity by drawing a parallel between both of them on the one hand, and between the objects of subject cataloguing, on the other.

Subject catalogues – classified as well as alphabetical ones – are, as we know, intended to satisfy those catalogue-users who ask for information about a given subject, without specifying the particular books or works which should contain this information. Accordingly, classified and subject catalogues provide a direct access, not to individual documents, but to definite themes; the organization of the information they contain is based directly on the themes, the subjects, the intellectual contents of the documents catalogued. This means that the structure of this class of catalogues accentuates the intellectual, and eclipses the material side of documents. One might say that it is the subject-contents of documents, that is to say, things of a purely intellectual, immaterial nature, which constitute the objects of subject cataloguing.

The structure of the author-title catalogue is more complex than that of all subject catalogues, in consequence of its having not only one, but three quite different functions.

As to the first function, many users looking for a particular book cannot help having in their mind the material side of the object they are looking for, seeing that in a book a particular subject-content is indissolubly connected with a material thing of a definite form. If the reader has seen the book he asks for, its material form will linger in his mind; he will expect to get a book bearing the very same physical marks which he remembers, and also anticipate that it will be retrievable by these physical marks in the catalogue.

Reversedly, the users in need of services pertaining to the scope of the second and third functions are concerned, first of all, with the intellectual side of the objects they seek. A catalogue-user looking for a work has no reason for being par-

[32] See § 43.

ticularly interested in the material side of the corresponding object of the second function, this material side being a batch of books, of which he wants to get, as a rule, only a single item. True, after consulting the catalogue he *may* develop a preference for a particular member of the batch, and thus his interest may be turned towards a particular material object. But in this case his aim has been transformed in the course of his consultation of the catalogue: originally a user asking for a particular work, he has turned into one asking for a particular book; the object he is now interested in, is no more an object of the second, but already one of the first function. However, many of the users in search of a work persist in remaining up to the end uninterested in the material side of the object they are looking for: for them all or several of the available editions continue to be entirely interchangeable, equally satisfactory. Part of this applies, *mutatis mutandis*, to the objects of the third function, too.

The objects of the second and third functions occupy, thus, an intermediate position between the objects of subject cataloguing and the objects of the first function of author-title cataloguing. They resemble the former in that the catalogue-user is not at all, or at the most only secondarily, interested in their material side. On the other hand, this circumstance does not induce him to revert to the subject catalogue: when wanting a particular work or author's 'oeuvre', he will try to trace it, just as when asking for a particular book, by its formal marks, i. e., in the author-title catalogue. In this instance, however, he will not be contented with being served with information only about those books which actually, visibly, display the marks he chooses to use as a handle, but will expect to find organized and full information on all the books which are components of that particular object of the second or third function which he is in need of. That is, he expects to get information on diverse books bearing different physical marks, neatly assembled under a single set of uniform marks; and this he does even if he has not been acquainted with that particular set of marks which the catalogue uses for this purpose. In this he resembles the users of subject catalogues. On the other hand, the close similarity between the outward forms of the physical and the uniform marks is another indication of the near kinship which connects the objects of the first function of the author-title catalogue with the objects of its two other functions.

31. THE RESPECTIVE MAIN SERVICES OF THE PHYSICAL AND UNIFORM MARKS. The cataloguer must, obviously, adapt his methods to the patrons' behaviour. From what was said in the former paragraph follows that he can accomplish this by making every book in its capacity as an object of the first function retrievable under its own physical marks; while in its capacity as a component part of an object of the second or third function he must make it accessible by means of the uniform marks of this object. This means – as this was mentioned in anticipation already in earlier chapters – that there exists a correlation between elemental objects (i. e., the first function) and physical marks on the one hand, and between composite objects (i. e., the other two functions) and uniform marks, on the other.[33] Let us

[33] S. LUBETZKY's *Cataloging Rules and Principles* (cf. our Note 3) contains a sentence which is flatly contradicted by our above proposition. "... While the second objective requires" – maintains Lubetzky on p. 38 – "that the works of an author should all be entered under one form of the author's

examine these correlations more closely; by doing so, we shall supplement our insight into the material or intellectual nature of the respective objects of the three functions with valuable details.[34]

We must begin by qualifying our former proposition according to which the indispensable means of recording composite objects, i. e., of the performance of the second and third functions, are the uniform marks, by pointing out that these means are indispensable only practically. Theoretically, it is not by any means impossible to perform these two functions without any use of uniform marks. As to the second function, one might achieve this by inserting into the catalogue for every work that has been published in several editions under several different sets of physical marks, after each single main or added entry or batch of entries headed by a particular set of these physical marks, 'see also'-references from this set to every other set under which main or added entries for other editions or variants of the same work are filed. This applies, *mutatis mutandis,* to authors' 'oeuvres' too. This solution would be equivalent to reiterating, though in part merely by referring to them summarily, all the information relating to a composite object under every set of physical marks under which one of the components of this object is re-

←

name, it does not dictate any particular form; whereas the first objective requires that the books should be entered under that form of the author's name under which he is bound to be cited and looked for most frequently by the users of the catalog". This passage has been, however, a mere slip of pen; this is made quite clear by the whole paragraph preceding it, in which it is repeatedly stated that the first objective calls for entry under the actual marks of books. For the fact, that Lubetzky's real opinion is expressed in these statements and not in the cited sentence, one may find as many proofs in his late writings as one wishes (e. g., in his Working paper prepared for the ICCP, cf. *Report of the ICCP*, London, IFLA, 1963, p. 140; or in his *Principles of Cataloging*, p. 93). The only reason for my having mentioned that passage is the regrettable fact that what with Lubetzky has been a mere slip of pen, seems to tend to become a doctrine in the eyes of some theorists, and survives even in these days. Cf. TAIT, JAMES A.: *Authors and Titles*. London, C. Bingley, 1969, p. 89.

[34] Against this one might object that, as the main objective of this study consists in investigating the functions of the catalogue, we ought to analyze the correlations not between the *two* classes of formal marks and the two most comprehensive classes of objects, but between the respective objects of the *three* functions and the *three* different devices needed for recording these objects. Such an objection would be, however, untenable. The most expedient device for discharging the second function – a recording of each elemental object under the uniform title of the work contained in it, and, if the latter's author is known, under the uniform name of this author – includes also the device indispensable to the fulfilment of the third function. From this it follows not only that in the present context we need not consider separately either these two functions, or the respective devices appropriate to their fulfilment, but also that we simply *must* not do this. The category relevant in this context is the comprehensive category 'uniform marks', and not the specific ones 'uniform name' and 'uniform title'.

It is certainly true that in cases when both the second and third functions become patent, for recording the three diverse objects of the three functions at least three different entries are needed: a main entry under uniform author's name and uniform title, an added entry or reference under uniform author's name and physical title, and a reference from the actual, the physical form of the author's name to the uniform one. But this is a fact we in this context are not concerned with, because the three different entries mentioned are the results merely of a particular – the optimal, the most expedient, but by far not the only possible – way of combining the application of the two main genera of formal marks. A cataloguer not choosing this mode can still fulfil the same functions, but only at a higher price: he will be obliged to use a greater number of entries (see text above). This shows that the choice between the diverse possibilities of combining the two main genera of formal marks is only a technical detail of secondary importance, dependent on the roles one chooses to assign to the three forms of entries which, I repeat, we are not concerned with. What we *are* concerned with, and what in this context is really significant, is in fact the problem of the correlations between the two comprehensive genera of objects (elemental and composite objects) on which the three functions between them are focused and between the two main genera of formal marks.

corded. It is obvious that this solution would be not only much less advantageous to the catalogue-user than the employment of uniform marks is, but also more laborious for the cataloguer, and more expensive for the library. So we may ignore this possibility entirely, and maintain that *practically* the second and third functions can be performed only by the application of uniform marks.

Uniform marks being, then, according to present-day notions and conditions indispensable to author-title cataloguing, logic requires us to raise the question whether it is not possible to simplify our cataloguing methods the other way round, by dropping the opposite device: the recording of objects under their physical marks?

The answer is decisively in the negative. The majority of those catalogue-users in need of a particular book who are acquainted with its outward appearance, its material constituent part will expect, as we have already said, to find it in the catalogue under the distinguishing marks of *this* constituent, i. e., under its own physical marks. The first function of the catalogue implies quite indisputably the categorical requirement that a catalogue-user of this type should not under any circumstances be allowed to go away empty-handed. His expectation of the proper location in the catalogue of the information he is in need of must be regarded as absolutely justified, and so the cataloguer must never fail in fulfilling this expectation. The librarian of the 16th century did not care to fulfil it and was in spite of this not entirely in the wrong. Since his time, however, the situation has changed fundamentally. In a large library of our days with its immense holdings, it is the author-title catalogue that provides the quickest access to a particular book, i. e., the shortest cut for the largest part of patrons; and it is the approach by physical marks which a majority of readers in fact tries to make use of. It is even more important that a great many among them are simply unable to try any other, since apart from its physical marks they do not know anything about the book they seek nor about its author.

Therefore, one of the main principles of author-title cataloguing is, obviously, that the catalogue should provide access to every document by its own physical marks, this being the cataloguing device the first function imperatively calls for. It should be stressed that – in spite of sporadic opinions to the contrary, e. g., some rules in the *Prussian Instructions*[35] – this principle is valid without any qualification and exception: the access by physical marks must be secured with absolute consistency. This holds even when the access to a particular book too, would be more efficiently secured by the uniform marks of its intellectual constituent; and, what is more, even when viewed from the side of the latter the physical marks of the book appear downright false and misleading. It may, of course, happen – though surely not frequently – that a particular book is in fact never sought under its physical marks. But this possibility does not absolve the cataloguer from acting with rigid consistency on the above principle, because he can never be certain of

[35] § 80 and § 183 of the *Prussian Instructions* provide for making additional entries under titles and author's names different from the uniform ones under which main entries are made, only if otherwise the access to an author or a "Schrift" does not seem to be secured adequately enough. In other words, this code explicitly declares that it is unnecessary to prepare additional entries under the physical marks of the books entered under a uniform mark, consistently and in any case. See: *Instruktionen für die alphabetischen Kataloge der preussischen Bibliotheken*. 2. Ausg. in der Fassung vom 10. August 1908. Berlin, Behrend, 1909.

its coming true. Moreover, there is also the even more forbidding generel argument against allowing exceptions, the argument that they always generate a certain uncertainty – let us recall that it is usually impossible neatly to separate the adjoining fields of application of two related rules –[36] that is in its turn liable to prepare the way for such neglect of the general rule which may already seriously impair the efficiency of the catalogue.

32. THE ACCESSORY SERVICES OF THE PHYSICAL AND UNIFORM MARKS PERFORMED IN EACH OTHER'S SPHERES. Physical marks are, then, indispensable to recording elemental objects, uniform marks to recording composite objects. With this statement, however, the theme of the respective correlations between the two genera of formal marks and these two top categories of cataloguing objects is not exhausted yet. Apart from their principal functions, both classes of formal marks perform also secondary services: both help to increase the efficiency of the catalogue also in the peculiar sphere of the other class.

An entry under the physical marks of a book discharges the first function satisfactorily only with regard to those catalogue-users who have seen that book, or have derived their information about it from a source stressing its physical marks. A by no means negligible portion of the catalogue-users looking for a particular book derive, however, the knowledge of its existence from sources which use in its description formal marks different from its actual physical marks. Such users will not find the book in the catalogue if it is not entered or referred to under these non-physical marks. That is to say, the first function can be performed much more satisfactorily, if every book is rendered traceable, not only by its own physical marks, but also by those marks of the work contained in it by which reference works and special literature generally refer to this work, in short, under the uniform marks linked with the latter. This is the way in which the use of uniform marks contributes to the efficiency with which the first function is fulfilled.

Reversedly, the performance of the second and third functions is also powerfully promoted by the recording of books under their physical marks. Not a few patrons in want of information on all available editions of a work or all available books of an author, do derive their knowledge about this work or author from an edition bearing physical marks different from the uniform marks of the work or from the uniform name of the author; or else from a source using such non-uniform marks. Though these users are in need of services which obviously fall within the scope of the second or the third function, a catalogue which secures access to the composite objects of these functions by their uniform marks only, but does not render them retrievable also by the physical marks of their components, will be of no help to them.

In short, recording books under their physical marks is not a device which serves exclusively the first function; neither is it the only device promoting the fulfilment of this function. The same also applies to the correlation between the uni-

[36] See p. 138.

form marks and the two other functions.[37] The secondary services of these two principal classes of formal marks generate also diverse classes of incidental objects of entries, which will be discussed, however, only in Appendix II.

For logical completeness, let us mention in parentheses that the catalogue can fail a user even if it fulfils all the requirements indicated above; the conditions of satisfying without exception all forthcoming reasonable demands are even harder. A book, a work, or an author's 'oeuvre' may be sought also under such arbitrary (i. e., non-physical) marks which differ from the uniform marks attached to them, but are used by a minority of literary or bibliographical sources. They may also be sought under physical marks which are different not only from the uniform marks attached to them, but also from the physical marks of all the pertinent books which the library happens to contain. Theoretically, the three functions cannot be completely fulfilled, therefore, but by registering all the variant sets of arbitrary marks attached to a composite object a component of which is contained in the library, along with the physical marks of all the components of that composite object, even of those that are not in the library. Surely, in this extreme formulation the claim seems rather absurd. Nevertheless, as to the variant names of a majority of authors and the variant titles of anonymous classics, in short, more or less in all cases when it is really important, this claim is in fact met by the catalogues of many large libraries, if not absolutely, but still fairly well. Differing on this point with L. Jolley,[38] I think this practice very sound. The inventiveness as well as the perseverance of the average reader leaves much to be desired, and therefore a cataloguer practising economy in the hope that a patron frustrated by the library catalogue will supplement his search by recourse to other bibliographical resources, is making a serious mistake.

33. THE CORRELATIONS BETWEEN THE DIFFERENT KINDS OF OBJECTS AND THOSE OF FORMAL MARKS. The above thesis that the correlation between elemental objects and physical marks, as well as that between composite objects and uniform marks, is not perfect, i. e., neither unequivocal nor exclusive, may be even stiffened by our pointing out that for the time being we are still unable to ascertain whether it is really a majority of the users for whom entries under physical marks procure the most direct access to a particular book, and entries under uniform marks to a work

[37] Here the objection might be raised that the lack of a complete correlation between functions and formal marks, pointed out in the text, is simply the outcome of our mode of defining the functions; and that the correlation might be made complete if only we chose to define the functions in another way: on the one hand, as a duty to meet the demands of those catalogue-users who seek an object of whatever sort (a particular book, work, or 'oeuvre') under its or under one of its components' physical marks, and on the other, as a duty to satisfy those who seek a work or author's 'oeuvre' under its uniform marks, or a book under the uniform marks of the work contained in it.

This is naturally true; but it would be an obvious mistake to define and to delimit the functions on the basis of the devices employed in discharging them. Our ultimate purpose is to illuminate the essence of the catalogue by means of clarifying the character of its main services. It is undoubtedly the objects about which information has to be conveyed under a particular function which impart to this function its character, its essence, – and not by any means the device, the particular genus of formal marks, used in conveying this information. So it is obvious that the definitions of the functions cannot be based on the latter, but only on the former.

[38] JOLLEY, L.: *The Principles of Cataloguing*. London, Crosby Lockwood, 1960, p. 11.

or an author's 'oeuvre'. In spite of this we maintain the proposition that the two categories of objects and formal marks *are* correlated in the sense that it is only between the elemental objects (i. e., the first function) and the physical marks as well as between the composite objects (i. e. the other two functions) and the uniform marks, that the relation is a logically necessary one. This becomes quite clear if we raise the question what we should feel obliged to do if forced to reduce the author-title catalogue to discharging only a single function and to applying only one of the two classes of formal marks. We can only conclude that in such a case the use of physical marks would necessarily be linked with the selection of the first function, and the use of uniform marks with that of the second or the third function. The service performed by either class of formal marks on behalf of the catalogue function(s) it is in this sense correlated with forms undoubtedly its intrinsic task, its very essence; and is at the same time the service most indispensable to the fulfilment of that (or those) function(s); while the service performed by either class of marks on the opposite side is only an accessory product of its discharging the former, its intrinsic, service; and is, though valuable, but not in the same sense indispensable as the latter.

Our findings concerning the correlations between the functions and the two classes of formal marks may be summed up in this way: The indispensable means of attaining the fulfilment of the first function is to make every elemental object retrievable by its own physical marks; that of the second, to secure access to every elemental object by the uniform marks of the work contained in it; and that of the third, to make every elemental object containing a work the author of which is known accessible by means of the uniform name of this author. The extent to which and the efficiency with which all the three functions are actually performed is, however, considerably increased by the recording of all elemental objects to which formal marks of both classes are attached, under two/three different sets of marks.

34. THE CUSTOMARY TREATMENT OF THE WORKS PUBLISHED IN COLLECTIONS IN THE LIGHT OF THE DUAL CHARACTER OF THE OBJECTS OF CATALOGUING. The last four paragraphs have clearly shown that with the objects of the first function it is their material side, their material constituent, with those of the two other functions their intellectual one, which is more important to, and has to be emphasized more strongly by, cataloguing. But it is necessary to stress once more that, in spite of this, the former objects are not of an exclusively material character, nor the latter of an exclusively intellectual, immaterial one. A great majority of the catalogue-users looking for a particular book are interested in the last analysis in its intellectual side, and also in the readers looking for a work or an author's 'oeuvre' an interest in the material aspect of this object is not wholly lacking. For the cataloguer, too, it is impossible entirely to ignore the less important side. There are not a few cataloguing rules, especially among the less fortunate ones, which make the choice of that particular one among its several diverse formal marks under which a certain elemental object should be entered, dependent on the subject-content of this object, for instance, the rules providing for a peculiar treatment of anthologies, or for entering corporate publications according to their dealing, or not dealing, with the activity of the issuing body. On the other hand, the rules deter-

mine the composition, the limits, of many a composite object of the second and third functions with a view to the material, the physical characteristics of their components. It is just this fact which has started us upon the preceding examination of the material or intellectual nature of the objects of cataloguing.

Now, the fact that the objects of the first function are by no means exclusively material objects, that they retain the compound nature which we have found to be a characteristic ascribed even by everyday language to books (and to all other documents, too), implies that the services of an entry - whether main or added – made under a physical mark of a book are not confined at all to conveying information merely about the material side of this book. This is quite self-evident with respect to books the intellectual content of which is constituted by a single work (even if there are secondary additions appended to it), and the physical marks of which conform entirely to the uniform marks of this work. But even in the absence of conformity of the two sets of marks, a catalogue-entry under the physical marks of a book, naturally, does not fail to communicate information also about the intellectual side of this book. This must be emphasized in view of the fact that present-day prevailing practice requires in such a case also an entry under uniform marks, the explicit and sole purpose of which is to communicate information about just this intellectual side. The existence of this entry does not alter the fact that the entry under physical marks, in spite of its primary purpose being quite a different one, does convey information about this same intellectual side too, although, of course, without emphasizing it, without relating it to the intellectual sides of other books, as the entry under uniform marks does.

We may mention here in parentheses that, in spite of this, it would not make any sense to stress this point to the extent of talking on these grounds about an inherent duality of the objects of catalogue entries, i. e., about each entry having two objects: a material and an intellectual one. This particular duality has no practical significance whatever, neither for the user of the catalogue nor for the cataloguer. Among the multiplications of the objects of cataloguing involved in the double-sided nature of books, only one is important in practice, and hence relevant to cataloguing theory: the multiplication entailed by the triplicity of the functions of the catalogue.

The lucidity of the cataloguing situation as well as the comforting adequateness of the generally applied cataloguing methods which we have met with in the case of books with simple contents, vanish at once if we come to collections, to books with complex contents. Here it is that we come across the curious phenomenon of the intellectual constituent of a book becoming in a certain sense obliterated by its material constituent,which has induced us to examine the question of the material or intellectual nature of the objects of cataloguing at such length. We have now reached a stage at which we may already attempt to explain and to size up this phenomenon.

As mentioned earlier, in the typical author-title catalogue most of the works contained in books with complex contents (e. g., the short stories contained in a one-volume anthology of prose) are suppressed; they do not appear among its autonomous objects. In spite of this, it would, naturally, be an error to presume that the catalogue records in such a case only a purely material phenomenon entirely irrespective of the intellectual phenomenon it is amalgamated with. The entry registers in this instance, too, a compound, two-faced object; but the intellectual constituent part of this object does not consist of the individual works the book contains, in their capacity as separate intellectual units, but of a new intellectual item different from all of these separate individual works: of the whole complex contents of the book. Books of this sort have usually a set of physical marks of their own referring to their complex contents, a set of comprehensive marks different from all sets of marks belonging to the individual works contained in them;[39] and they accentuate only this comprehensive set of marks, or accentuate it at least more strongly than all the respective sets of marks of the works which constitute their contents. In consequence, as to such books the first function calls – according to what we have said about the correlation between this function and the physical marks – first of all for a record under this comprehensive set of physical marks. This is one of the causes which have induced cataloguers to content themselves with recording such books under their comprehensive set of physical marks only, and to refrain from recording also the intellectually separate constituents of their compound contents, from making these constituents, too, autonomous objects of cataloguing.

It cannot be denied, however, that this practice permits the material constituent of a book in a certain sense to eclipse its intellectual one, and that it involves also an odd inconsistency in the selection of the autonomously recorded objects. The principle from which this practice originates is the very sound one that it must be secured in any case that all physically separate units should be rendered retrievable by means of their own physical marks. One of the outcomes of this practice is, however, the at first rather astounding fact, that our author-title catalogues never omit to register a *physically* separate item, but fail to record very many *intellectually* separate, self-contained and independent items, with a lot of highly significant and frequently demanded ones among them.

Is it possible to justify the second part of this practice?

The most obvious argument which may be raised in favour of it is of an economic nature. If we aimed at recording every intellectually separate and independent work contained in books with complex contents, this would entail preparing separate entries for every poem contained in volumes of poetry, for every paper read at congresses, etc. Under present conditions this is not really feasible; the increase in the costs and the clumsiness of the resulting catalogues would be forbidding, and certainly not balanced by a corresponding increase in their efficiency. Moreover, the intellectually independent items in question are rather often not equipped with formal marks suitable enough to be used as headings of entries (e. g., the documents in a cartulary). We must, nevertheless, confess that all these arguments are not quite conclusive. It is not at all impossible that the librarians of a not very distant future may turn up their noses at this practice of ours.

The habitual neglect of recording autonomously the physically not separate constituents of complex book-contents is, however, as everybody knows, only partial. The range of the types of books which come under this head is very wide; it stretches from journals to chrestomathies, from encyclopedias to the *Bible*. In regard to a lyric anthology or a cartulary most present-day cataloguers accept the

[39] If they have not, cataloguers are used to treating the physical marks of the work first mentioned on the title page or occupying the first place in the book, as if they were the physical marks of the book itself – just as though the latter did not contain any further works.

above arguments as a sufficient justification of the practice under consideration; confronted with an anthology containing five or six plays, or with the proceedings of a symposium consisting of five or six lectures, the same cataloguers are likely to become already doubtful and less at ease when they omit to record autonomously each single play or lecture. However, when coming across a book with the title *I Quattro Poeti Italiani,* containing the *Divina Commedia,* the poems of Petrarca, *Orlando Furioso* and *La Gerusalemme Liberata,* most of them will feel compelled to transgress the limit drawn by the criterion of physical separateness. The urge to record, apart from the book with complex contents as a whole, also the single works contained in it increases, obviously, in direct proportion to the excellence, and in inverse proportion to the number of the latter. This urge is the source of analytical cataloguing in the narrower sense. The objects added by analytical cataloguing to the stock of the objects of cataloguing do not belong, however, to that class of objects we are at present concerned with and which may be labelled as 'books'; we shall deal with them in another context at a later stage of this study.

35. Terminological conclusions. We have arrived now at the end of our analysis of the meaning which is to be assigned to the term 'book' in the definition of the first function. In spite of all its complexity, the meaning delineated above is, I think, in all essentials in accord with the opinion on the scope of the first function prevailing among cataloguers, and also with the intended meaning of the note in the *Paris Statement of Principles* according to which the word 'book' is to be understood to comprehend all "other library materials having similar characteristics".

Before concluding this analysis, let us draw a few terminological conclusions. To designate that class of objects of the first function which we are at present discussing, even in our own days most authors still keep using, instead of the word 'book', the word 'work'. This is, since misleadingly unprecise, impermissible. In cataloguing terminology the latter term should be reserved for the purely intellectual phenomenon which is only one of the two constituents of the double-sided phenomenon 'book'; and it is evidently absolutely impermissible to use the term in two different senses. To demonstrate this, it suffices to point at two facts: first, that there are several sorts of books the subject-contents of which are certainly not included in the everyday meaning of the word 'work' – this topic will be taken up in a later chapter;[40] and secondly, at the fact just discussed, that in distinction from books, a considerable part of the works which undoubtedly *are* included in the everyday meaning of this word do not become objects of cataloguing even if they get into the hand of a cataloguer. Such an improper use of the word 'work' is not only logically untenable, but also detrimental to practice, to the clarity of the rules of codes, as it will be shown in another context (cf. § 55).

The use of the word 'book' does not lack in problems either. If we do not wish to refrain from using it in its everyday sense, then it is, naturally, indispensable that we should abstain from using it in the above outlined technical sense, replacing it in this capacity by another word, a special technical term. For our part, if in the following we shall be in need of such a term, i. e., of a designation for the main class of objects of the first function, we shall use instead of the word 'book' the

[40] See p. 91–92.

term 'primary elemental object' of cataloguing. The adjective 'elemental' serves to contrast the objects of the first function with the composite objects of the other two functions. The adjective 'primary' serves the purpose of contrasting that sub-class of elemental objects which *is* to be comprised in this technical term, with those two sub-classes which are *not* (the latter will be surveyed in the next chapter).

Against our new term the objection might be raised that we possess already a term, the term 'bibliographical unit', which though used in widely diverging various senses, has also a prevailing denotation. This seems in all essentials to be congruous with the denotation of our term 'primary elemental object', though I have not yet come across an exact delimitation of it. There is, however, a weighty argument against our adopting this term. An outstanding feature of the term proposed by us is that it lays explicit stress on the circumstance that the objects covered by it form only a part of the whole bulk of the objects of the first function, which includes along with them also other classes of objects. This is an outstanding and indispensable feature of the term we need. The term 'bibliographical unit', however, is always used as a comprehensive term embracing *all* the objects of the first function; it implies the assumption that besides the books in the extended sense of the word outlined above there are no other objects within the range of this function. In other words, it has an ingrained connotation entirely incompatible with either the facts, or the terminological needs of theory. So its adaptation for our purposes would unavoidably turn out to be confusing and unserviceable.

36. PRACTICAL CONCLUSIONS. (a) CODES ARE NOT ARTICULATE ENOUGH IN REGARD EITHER TO THE ROLE OF PHYSICAL SEPARATENESS OR TO THE APPROPRIATE HANDLING OF BOOKS AMALGAMATING SEVERAL PRIMARY ELEMENTAL OBJECTS. As to the fundamental point, that the cataloguer must not fail to render every primary elemental object retrievable, i. e., that he must record every single such object autonomously, there is no disagreement among cataloguers. The codes, however, fail to express this cardinal principle with an appropriate articulateness. The wording of many rules providing for the cataloguing of anthologies, for instance, prescribes that they should be entered under the compiler, with an added entry under the title. This implies that the two editions of the anthology of short stories of our above example have to be catalogued in exactly the same manner, irrespective of their having been published once in one volume, and another time in ten volumes. The rules alluded to exclude, literally taken, the possibility of a separate, autonomous recording of the single short stories in the latter case. This is, obviously, not in harmony with the real intentions of the compilers of the codes in question, and so one has to regard the wording of those rules as simply incorrect. The mistake originates in these codes having entirely neglected to state in general that to the physical separateness of the items catalogued they intend to assign a crucial importance and a decisive role.

Another omission – of an even more general and material consequence – is that codes fail explicitly to state that in many single physical objects there are two or more primary elemental objects united, all of which must, on principle, be recorded separately, in their own right, and under their own formal marks. Most codes provide in one set of rules for the cataloguing of single books, and in another set for the cataloguing of series, without mentioning that in general the latter set is not to be applied without employing at the same time to the very same physical items also the former set.

A catalogue omitting to secure access to a particular volume of a series in either of these ways, neglects one of its most prominent duties and is heavily defective. This is an extremely important fact to which the code-maker must not fail to give explicit expression. Here is one of the points at which he has to be reminded that his business is not tacitly to understand things, but to state clearly and definitely what he expects the cataloguer to do. To this statement he must give a general form and a place among the fundamental rules of his code. It cannot be substituted by casual – and unavoidably defective – provisions concerning the preparation of added entries in one or the other special case of an amalgamation of two kinds of primary elemental objects in a single physical item.

37. PRACTICAL CONCLUSIONS. (b) THE CURRENT TERMS FOR THE DIVERSE TYPES OF PUBLICATIONS HAVE NO CLEAR-CUT DENOTATIONS. Another shortcoming common with most present-day codes consists in their basing the rules for the treatment of the more comprehensive one of the two primary elemental objects united in a single physical item, upon terms the meaning of which is very far from being unequivocal. For instance, they often give two divergent sets of directions for the cataloguing of series and of collections, just as if the line dividing the denotations of the two terms, and the two classes of publications these denotations include, were precisely and clearly agreed upon – which it is very far from being. The same applies to the boundary lines separating either of these two types of publications from the single books published in several volumes (which, of course, do *not* unite two different primary elemental objects of cataloguing).

Other terms belonging to the same family and habitually being used, must be objected to on the grounds that they subsume widely differing cataloguing situations under the same head. In some codes, for instance, the word 'collection', and, of course, also the rules providing for the treatment of collections, embrace a broad spectrum of widely different publications, beginning with a one-volume lyrical anthology and ending with *Muratori's Rerum Italicarum Scriptores*. That is to say, these codes throw items ranging from a single simple object of cataloguing to a monumental and extremely complex publication, almost every physically separate part of which unites at least two and very often several primary elemental objects, into the same hat.

Undoubtedly, logical and terminological untidinesses of these kinds are shortcomings from which codes ought to be purged, not only for the sake of logic, but also for practical reasons. As we have already stressed, a code must be articulate, and has no business to assume that cataloguers know, without being told explicitly, what they are expected to do. The fact that even an almost senseless rule will not prevent a majority of cataloguers, relying on tradition and usage usually much older than the code, from supplementing the sense and finding the right course, is no excuse at all for tolerating any logical or terminological laxities in our codes. It is indeed deplorable if there are instances in which catalogues can become adequate only *in spite of* the code, by means of cataloguers treating the code as a negligible quantity. One of the foremost duties of a code is to give instruction and help to cataloguers not yet competent enough as well as to students, but an obscure code is liable to fail just these among its users.

The terminological untidinesses in question might be diminished by fixing with the highest attainable degree of precision the denotation of the terms 'series', 'collection', 'periodical' etc., combined with an examination of the catalogical relevance and utility of these terms, resulting in the discarding of useless, and in the introduction of wanting distinctions, in short, by an adaptation of our terminology to the real needs of cataloguing theory and practice.[41] We are not unaware of the fact that it is not always possible to fulfil the first requirement to perfection. This is, indeed, a somewhat discouraging fact. It implies that we cannot succeed in determining the exact limits even of the range of primary elemental objects, in other words, that an attempt at preparing entirely exhaustive, gapless cataloguing rules is doomed to fail already at the first step: at the fixing of the limits of the principal and most important class of the objects of cataloguing. The cause of this is that we have to deal here with one of the by no means rare points, at which the immense variety of the forms, greatly aggravated by the abundance of intermediate and mixed forms, with which the authors, compilers, editors, publishers and printers insist to fit out the individual books, inevitably thwarts all efforts to construct fully gapless, all-embracing cataloguing instructions. For closing the gaps which for this reason cannot be eliminated from our codes, we have no other recourse than to the catalogue-worker's sound judgement. It is, among others, his exercise of judgement in case he turns in vain to the code for directives, on which the acknowledgement of the intellectual character of his performance has to be based, not on his occasional 'disregarding of codes and rules'. But in regard to the problem under consideration we are still far from the point where there is no other way out than a surrender of the field to the catalogue-workers' discretion. By a narrowing of their terminological inadequacies, the rules we are here speaking about can still be improved considerably.

38. PRACTICAL CONCLUSIONS. (c) ARGUMENTS IN FAVOUR OF A GENERAL APPLICATION OF DETAILED SERIES ENTRIES. But it is not only the habitual formulation of the rules concerned with tasks in themselves unquestioned, which leaves in this context still much to be desired; we must also point out a task which is, as compared with its real significance, heavily underrated.

Regarding an issue of a periodical, or a volume of a multi-volumed complex publication, devoted to a self-contained subject and equipped, apart from the marks of the complex publication, also with separate formal marks of its own, there is no disagreement: no respectable cataloguer would ever think of shirking the autonomous recording of either of the two primary elemental objects united in a physical unit of these kinds. Far less satisfactory is the situation with regard to common series. The *Prussian Instructions* are frankly averse to the detailed registration of a publisher's series. They prescribe, in general, only a recording of its constituent volumes under their own respective formal marks, i. e., in their capacity as single detached books, but not also under the formal marks of the series, in

[41] It is, for instance, obvious that the term 'serials', however useful in visible indexing, is rather out of place, and even confusing, in the terminology of author-title cataloguing.

their capacity as the latter's parts. Although the big codes of a later date have already abandoned this policy in a rather marked degree, some of the codes for public and little libraries are sticking to it unwaveringly even in these days. In my opinion, this is a mistake, at least as far as card catalogues using unit cards are concerned, with which the costs of detailed series entries are in comparison to their usefulness infinitesimal. After all, the series title is the most efficient one among all the available auxiliary handles, all the supplementary access points, which one may establish for the benefit of users who have forgotten the name of the author or editor, as well as the correct title of the book they are looking for. But even the librarian himself, the book-buyer as well as the cataloguer, will often find detailed information on series extremely useful. Moreover, the detailed registration of a series with a limited range of subjects can also perform a further service: it may be appreciated by patrons as an additional means of subject approach, and a very easily accessible one at that. In public libraries the last argument carries, obviously, even more weight than in large research libraries.

Summing up, I think it desirable that the preparation of detailed series entries should be generally recognized as an integral part of the first function, to be performed, most conveniently by means of series added entries, by one at least of the author-title catalogues in every library.

Accordingly, I regard § 33N of *AACR* as a very happy rule, inclusively of the exception of those series which might not reasonably be catalogued as a collected set. Much less happy seem to me the further exceptions the rule makes, concerning those 'works' that cannot be reasonably assumed to be cited as part of the series, and those series the title of which includes the name of a trade publisher, or which do not have a subject limitation and have only format in common.

CHAPTER V

THE FIRST FUNCTION
II: ELEMENTAL OBJECTS OTHER THAN 'BOOKS'

39. CORRELATIONS BETWEEN THE CATALOGUE-ENTRIES AND THE OBJECTS OF CATALOGUING. THE IRRELEVANCE TO US OF THE FORM OF ENTRY. In the preceding chapter we have specified the denotation which – on the basis of actual world-wide cataloguing practice – one must assign to the word 'book' in the current definition of the first function; and, since this denotation differs very considerably from its everyday meaning, we have proposed to substitute for this word, when a special designation for the chief objects of the first function is needed, the term 'primary elemental object of cataloguing'. We have also mentioned in passing that there are several kinds of objects belonging to the domain of the first function which are not only no'books', but which are not included in the denotation of the term 'primary elemental object' either. Our next task consists of making an inventory of the objects belonging to this category: i. e., of the objects which do belong to the domain of the first function and so to the class of elemental objects, but *not* to its sub-class 'primary elemental objects'.

We have told above that all elemental objects are, by definition, autonomous, which means that there must be at least one catalogue entry devoted to, and headed by a formal mark of every such object. We can make use of this fact in solving our present task: we propose to conduct the search for non-primary elemental objects by surveying all the different kinds of entries applied in author-title cataloguing, and after setting aside those devoted to primary elemental objects, by examining the rest and classifying it according to their objects.

Preliminarily we must emphasize a point that is important, not only to the survey presently to be undertaken, but also to our inquiry as a whole. This point is that the three generally applied, functionally as well as formally different forms of catalogue entries (main entry, added entry and reference)[42] are from our point of view completely equivalent: it is of no relevance at all which of them is used for recording a particular class or kind of objects. Only one requirement is essential; all the objects which are intended to be made autonomous should be rendered accessible by at least one entry by means of their own formal marks but it is almost irrelevant whether that entry is a main entry, an added entry or a reference. At first

[42] We did not include in this specification of functionally different forms of entry the analytical entry in the traditional, narrower sense of the word, on the grounds that it has no specific function of its own, different from the functions of non-analytical main entries.

glance, this proposition seems to be irreconcilable with the fact that the choice of the main entry heading still belongs, with regard to quite a number of cataloguing conditions, to the most controversial issues of cataloguing theory and policy. The apparent contradiction can, however, be easily resolved. That the choice of the appropriate form of entry, and particularly that of the main entry heading, should be a happy one, is indeed of the utmost importance but in the logical sequence of the topics with which the theory of cataloguing must deal, it is necessarily preceded by a general survey of the objects of cataloguing. In this study we are concerned with the latter theme only (more precisely, with one section of this theme: we have to take stock of all – actual as well as desirable – kinds of autonomous objects of cataloguing within the range of standard cataloguing), while the choice of the form of entry fittest for recording a particular kind of object is a topic which forms a later chapter of cataloguing theory, and lies thus outside the scope of this inquiry. Though we shall touch upon this topic now and again, we shall do it only incidentally, in connection with the practical conclusions which we shall occasionally draw from our argument proper, but never for its own sake. The topic of the correlations between the diverse sorts of entries and objects we shall, however, take up at some length in Appendix II.

Let us now illuminate more closely the proposition about the equivalence of the three forms of entry from our point of view, and develop it in greater detail.

So far as the equivalence of the main with the added entry is concerned, the validity of this proposition is obvious enough. A cataloguer, entering e. g. an edition of Calvin's *Commentary to the Psalms* containing also the complete text of the latter, who allots the main entry to the *Commentary* and only an added entry to the *Psalms* (or the *Bible),* agrees upon the autonomous objects of cataloguing completely with that other cataloguer who applies the two forms of entry in the reverse manner. Both of them intend to record the very same two distinct elemental objects; they agree also in wishing to economize a second main entry, i. e. to allot the status of a *primary* elemental object only to one among the two objects; and differ only in a single, for us irrelevant matter, namely in their choice of the object which shall pay for this economy.

Less obvious is the equivalence to a main or added entry of a reference in cases where there is no access to an elemental object except by this reference. This is the case whenever a main or added entry which conveys information – not repeated in any other entry of the catalogue – about an elemental object, is headed not by a physical mark of this object, but by a uniform mark. In such a case even the fact of this elemental object being recorded at all becomes obscured. Let us take, for instance, the references advising catalogue-users to look under the uniform name of a particular author also for all those books of his which have been edited under another name. These books, i. e., a lot of diverse primary elemental objects, are entered under the uniform name, and under the name they actually bear only a single name-reference is given. Their main entries refer to these books only indirectly, in their capacity as component of a composite object. It is the uniform mark of this composite object which is used as the only handle facilitating the retrieval of the information about them; while such entries which would record them as independent, separate elemental objects, under their own respective physical marks, the catalogue is entirely lacking.

As to this practice it is to be stressed that we must not let ourselves be lured by the outward appearances it entails into thinking that it involves a neglect of the duty to record books in their capacity as elemental objects, in other words, a neglect of the first function. The reference from the name it actually displays to the uniform name, renders each of the books covered traceable also under its own physical mark; it is in fact a substitute for as many separate entries serving the purposes of the first function as many books (and other elemental objects) are in the library that display the name referred from. It is a device for the simultaneous recording of several diverse elemental objects, and a very rational and ingenious one; besides, as its early appearance and application show, also a very plausible one. The purpose it serves is purely economic: to avoid a duplication of a whole batch of entries, by substituting in one of two instances for the batch a single comprehensive, though less efficacious entry.

Notwithstanding this, one might easily feel inclined to regard the form of entry richest in information – i. e. the main entry – as the only instrument fit to record *primary* elemental objects appropriately, that is, under the formal marks most closely connected with them: their physical marks; and to consider, therefore, as arbitrary a practice which relegates this task to the form of entry poorest in information. With this misgiving we shall be able to deal only at a later stage.[43]

40. ALL ENTRIES HAVE SEVERAL OBJECTS. THEIR OBJECT PROPER AND THEIR COMPLEX OBJECT PROPER. Preliminarily to our search for non-primary elemental objects we must emphasize that it would be a mistake to believe that the application of the method of search suggested above will be an easy job. It could be so if the correlations between entries and objects were always clear and unequivocal, which would, however, be the case only if every entry had merely a single object, and every object a single entry; or more correctly, if the function of every entry were confined to serving the retrieval of a single object, and the retrievability of every object depended on a single entry only. As a matter of fact, neither of these two conditions is fulfilled. So we had better clarify first all those correlations between the entries and objects of cataloguing which are relevant to our search, but not clear enough in consequence of the blurring effect of other correlations with which they are intervowen.

As repeatedly emphasized, we are interested only in autonomous objects. An object the retrievability of which is not served by any entry having one of its distinguishing marks for heading, does not concern us at all.

Let us examine first the difficulty involved in the fact that every entry promotes the retrieval not only of a single, but of several autonomous objects. Apart from the object whose formal mark it bears for its heading, and whose autonomous character it is, thereby, establishing, every entry helps to facilitate also the retrieval of at least one further object the autonomous status of which depends not on it, but on another entry. There is indeed not one entry in our catalogues which does not secondarily serve in a greater or lesser degree to enhance the retrievability of one or more elemental and/or composite objects of this description. According to these latter objects of theirs, entries may be divided into classes and sub-classes,

[43] See § 61.

some of which serve the retrieval of assortments of objects which include also non-primary elemental objects, while some others do not. In the present context, naturally, only the former are of interest.

Among the different objects of an entry that particular object which it directly refers to (i. e., a formal mark of which it has for its heading) is to be regarded as its *principal* object. Not being concerned but with autonomous objects, we can confine our search for non-primary elemental objects to these principal objects of entries only, seeing that each elemental object is also autonomous, and must form therefore the principal object of at least one entry. We may ignore all non-principal, all incidental objects of entries, without incurring the danger of overlooking any elemental objects.

At this point, however, we are confronted with a complication. Strictly taken, the above definition of the principal object implies that an entry with a physical mark for heading must have an elemental object for its principal object; while the principal object of an entry with a uniform mark for heading must be a component of one or more composite objects, it must be one or more intellectual aspects of the elemental object concerned – but not this elemental object itself. So, being concerned with elemental objects only, at a first glance we might be inclined to infer that we may confine our search to the entries headed by a physical mark, making thus a further step in narrowing down the range of entries we shall have to review and to sift.

Such a conclusion would be, however, erroneous; it would mislead us into overlooking quite a considerable portion of elemental objects, primary as well as non-primary ones. As this was already touched upon in the foregoing paragraph, in consequence of the triplicity of the catalogue-functions and the cataloguing techniques originating from it, the physical marks of many elemental objects do not appear except in the headings of references which have for their principal objects mostly whole batches of elemental objects, and which do not convey about the single elemental objects these batches comprise the necessary amount of information. The indispensable characteristics of an in itself self-contained entry – its enabling the catalogue-user, first, to locate, and second, to identify in the catalogue the object he is looking for – are distributed in this case between two separate entries: between the reference pointing from a physical mark of an elemental object to the corresponding uniform mark, and the main or added entry prepared under the latter, which alone conveys adequate information about that particular elemental object. When setting out to obtain a complete inventory of elemental objects, we must not omit to take into consideration this class of main and added entries either, in spite of their being headed by uniform marks.

The main and added entries headed by a uniform mark have, then, along with their principal object which bears the character of a component of composite objects, another object of an even greater importance to our subject: the elemental object, also which is the reverse of their principal object: the component of composite objects. This implies that we are faced with the necessity of distinguishing between, and of employing the concepts of, two different kinds of principal entry objects: first, the elemental object a formal mark of which forms the heading of the entry, without regard to the circumstance whether it is the physical or the uniform variant of this mark which actually appears in that heading; and second, the composite object – or more precisely, the component of (one or more) composite objects – which the uniform mark in the heading directly points at.

We are also in need of two different terms for these two differently defined principal objects. We propose to apply to the former the term 'object proper', to the latter the term 'complex object proper' – in spite of the fact that it would appear more logical to assign the simpler designation to those objects which we were calling until now simply 'principal'. With an explanation for our preferring the reversed terminology, as well as with a more detailed discussion of the necessity to employ two different concepts of, and terms for, the principal objects of entries, we do not wish to encumber our present train of thought, and relegate this detailed discussion to an appendix (cf. Appendix II., pp. 149–151).

Summing up our findings, we state that it is possible to narrow down our search for non-primary elemental objects and thus to simplify it quite radically. First of all, we need not consider main entries, since their object proper is always a primary elemental object. Nor need we consider references which, because of the scarcity of the information they contain, cannot give us any help. So we may confine our attention to added entries. But of these, too, we may exclude a particular class: the added entries prepared on account of two (or more) primary elemental objects being amalgamated in a single book; these we may disregard, because all the elemental objects involved are *primary* ones. On the other hand – let us emphasize this once again – we must extend our scrutiny also to the added entries with a uniform mark for heading.

41. CLASSIFICATION OF ADDED ENTRIES ACCORDING TO THEIR OBJECT PROPER. CO-ORDINATE ADDED ENTRIES. The last possibility of diminishing the field we have to search for non-primary elemental objects consists in excluding another class of added entries, which, however, we must first delimit and define.

The added entries prepared in the course of cataloguing books not uniting two different primary elemental objects, may be divided into two main classes: the object proper of those in the first class is identical with the object proper of the pertaining main entry, whereas the added entries of the other class have an object proper different from that of the main entry. The entries in the former class are, obviously, of no relevance to our search; but just for this reason we must try with every possible care to fix the boundary line dividing the two classes.

Let us begin with referring to the well-known fact that in standard cataloguing it is the name of the personal author which is used as the primary formal distinguishing mark of the objects of cataloguing. When the author of the work constituting the content of a book is known, the majority of cataloguers content themselves with preparing a single entry, an entry under his name, except when this does not seem to be a satisfactory enough handle. This is the case, first, when the author's name is not mentioned at all or is not sufficiently emphasized in the book – in these cases we might speak of 'hidden authorship' – and second, when, by an arbitrary extension of the concept 'author', the author's rank, as well as the leading role of his name among the formal distinguishing marks of books and works, are transferred to the compiler or editor of a collection, or else to a corporate body – these cases we may subsume under the term 'pseudo-authorship'. In all of these instances every conscientious cataloguer considers it his duty to improve the chances of the catalogue-user by preparing an added entry under the title.

If only a single entry is made for a book, a main entry either under the author's name or (in case it contains an anonymous work) under the title of the work contained in it, there cannot be any doubt as to the object proper of this entry: this cannot be anything else but the two-faced phenomenon 'book' itself which is identified by the heading together with some other data contained in the body of the entry. This is quite clear, seeing that it is the book itself which the cataloguer is under any circumstances and first of all obliged to record. Less obviously, but still indisputably, the object proper of an added title entry accompanying a main entry made under the name of a hidden or a pseudo-author cannot be anything else than the same book which also forms the object proper of the main entry. The only reason for making such an added entry is, from the point of view of the first function, the comparative unsatisfactoriness of the approach provided by the name of a hidden or pseudo-author, and not by any means the emergence of another, additional elemental object calling for being recorded in its own right. In comparison with these two kinds of a not entirely adequate author-access, the title-access is generally not inferior; it would be, therefore, an arbitrary course of proceeding for the cataloguer to content himself with a single entry made under the name of the hidden- or the pseudo-author.

Apart from hidden and pseudo-authorship there are many similar instances of two – or even more – parallel entries being required for recording a single primary elemental object adequately. The necessity for this kind of multiplication of entries emerges, as a rule, when a primary elemental object is equipped with two or more formal marks which essentially differ from each other in their denotation, but only slightly in their suitability to identify and to locate this object. The most notable instances of this situation are the following: editions of works by two or three co-authors; a large part of the publications of corporate bodies, to be recorded under their title as well as under the name of the body, and often also under the name of a personal author, editor or compiler; publications with complex contents, as well as a great many editions of anonymous works naming a compiler or editor; anonymous publications carrying two different titles, or a subtitle formally as strongly emphasized as the main title, or more distinctive than the latter; thoroughly revised editions of a work, if the reviser is a person other than the author; books containing together with the text of a work also an exceptionally notable commentary to it written by a person different from the author of the original work; editions of biographies, an essential part of which is formed by excerpts taken from the works of the person represented, interwoven with texts written by the biographer; editions of reports of interviews made with one or two, or of conversations between two or three, partners; editions of works written by a person other than the conveyer of the information underlying it; early dissertations naming a *praeses* as well as a respondent; editions of only a part of, or of a continuation or supplement to, a work or a collection published and catalogued usually as if it were a single integral work (e. g., the *Bible*), having also a separate title or author of its own; festschriften; etc.

It is current world-wide practice to enter primary elemental objects pertaining to these types under both or all their formal marks approximately equally suited to facilitate their retrieval. The two or more entries involved in any of these cases relate to the very same single primary elemental object; apart from a single main entry in each case, they are added entries (or added-entry-functioned references). The added entries of this class are in function, efficiency and importance approximately equal to the main entry they are attached to. For this reason we propose to designate them as 'co-ordinate added entries'; and repeat our former statement now in the form that co-ordinate added entries are not of interest to our present search.

42. SECONDARY ELEMENTAL OBJECTS AND ACCESSORY ADDED ENTRIES. With this we have, at last, finished sorting out all those classes of entries which cannot put us on the track of a non-primary elemental object. The residue left over after this sifting is relevant to our survey in its entirety: all the remaining entries – which are without exception added entries – do have non-primary elemental objects for their object proper.

The residue in question divides into two main groups. To the more important group belong, first, the added entries the object proper of which is one among several intellectually separate and self-contained items of equal standing that constitute the complex contents of a primary elemental object, e. g., an individual paper contained in the report of a symposium, and second, the added entries recording subordinate constituents of the contents of primary elemental objects, e. g., appendices, accompanying studies, introductions, prefaces, epilogues, commentaries, or other explanatory material, the author of which is not identical with the author under the name of whom the primary elemental object is entered. Illustrations, photographs, and other accompanying matter belong also to this class of secondary constituent parts of book-contents.

A parenthetical terminological remark. To the term 'analytical cataloguing' there is sometimes attached a limited meaning which includes only the items belonging to the first point of this specification. It is, however, obvious that the items which come under the second point are closely related to the former, and that, therefore, it is quite plausible to invest the terms 'analytical cataloguing' and 'analytical entry' with more comprehensive denotations, embracing *all* the above enumerated kinds of component parts of book-contents.[44]

The function of all the added entries specified under the above two points is, without doubt, essentially different from the function of co-ordinate added en-

[44] The topic of these two different classes and concepts of analytical entries will be examined more closely in Appendix II. To the earliest instances of analytical cataloguing belong those books with complex contents in which the works constituting these contents are preceded by inner title-pages displaying only the respective formal marks of these works, but not also the marks of the book as a whole. The reason for these works having been promoted so soon to the rank of autonomous objects was evidently the intermediate nature of the formal characteristics of the books that contained them. Early cataloguers felt the title-page to be a sign of separateness and independence, and this induced them to treat even items endowed only with an *inner* title-page, although they were quite evidently merely the constituents of a book, as if they had been separate and independent books. Our predecessors were, of course, quite right in doing so, seeing that readers were likely to react to the inner title-page in exactly the same manner.

A collective volume is, naturally, essentially different from a book with an inner title-page. The best way of treating books united in such a volume is to catalogue them in just the same manner as if they had retained their original physical separateness and independence, and to indicate their condition in a note or within the location mark only. Consequently, for us collective volumes do not raise any special problems whatsoever.

tries: it is not confined to procuring an additional access to the object proper of the main entry, but includes also the recording of another object, quite different from the latter. The object proper of a co-ordinate added entry is always a primary elemental object; such an entry refers to this object as a whole, and is headed by a formal mark of this very object. On the other hand, an added entry recording a particular piece published in an anthology, or else an introductory study, is not made in the first instance for the sake of improving the retrievability of the primary elemental object in the course of the cataloguing of which it has been prepared, but serves in an at least equal degree the purpose of securing an access to that particular piece in the latter's contents. Accordingly, it is headed by one of the formal marks of this piece, which differ wholly or partly from the formal marks of the primary elemental object in which the piece is contained. It is obvious that added entries of this kind increase the bulk of the information imparted by the catalogue; they enrich it by items referring to otherwise not recorded elemental objects. As to their intellectual or informational value, these new objects are often manifestly equal to many primary elemental objects; e. g., an analytically catalogued short story contained in an anthology is in a high degree interchangeable with, and equivalent to, the separate editions of the same short story. That an object of this kind is, first, never a physically separate publication, and second, that it is always assigned an added entry only – these two circumstances do not affect in the least the fact of its being a distinct and separate elemental object of cataloguing in its own right.

The elemental objects of this kind, which no variant of the current definition of the first function ever mentioned, we propose to denominate as 'secondary elemental objects' of cataloguing; while to the added entries relating to such objects we shall apply the designation 'accessory added entry'.

43. Tertiary elemental objects and tertiary added entries. The accessory added entries do not entirely absorb our residue of added entries having a non-primary elemental object for their object proper. We have as yet not spoken of the entries made under the name of a translator or editor of a book or secondary elemental object entered under personal author, or else under the name of a corporate body which played a subordinate but still intellectual role in the production of a book or secondary elemental object recorded by means of a main or added entry headed by a title, the name of a body with a role of greater consequence or, in a single specific case, the name of a personal author. Bodies with such subordinate roles are the bodies with whom a translation originated, the sponsoring bodies, the bodies which are the hosts of congresses, conferences, or symposia, and in some cases even the editing bodies. All the enumerated entries, the object proper of which consists in some kind of personal or corporate participation of an intellectual character in the production of a book or work, in other words, in a specific aspect of this book or work, form, in our opinion, a distinctly separate class of added entries. The reasons for this opinion are the following.

The added entries belonging to this group cannot be classed either as co-ordinate or accessory ones. They conform with co-ordinate, and sharply differ from accessory added entries in referring *not* to a particular, intellectually as well as physically distinguishable separate *part*, a section of the contents of a primary elemental object only. Their congruity with co-ordinate added entries ends, however, at this point; they differ from the latter too, as their efficiency in facilitating the retrieval of the primary elemental object they are related to is by far not comparable with that of the main entry. In this of its capacities the value of an added entry, e. g., under a translator or editor of a personal author's work is almost negligible. So, the function proper of these added entries cannot be considered to consist in providing an additional access point, equivalent to the main entry, to the respective primary elemental objects, but only in recording a particular aspect, a particular feature of these objects, the feature of their having been translated or edited by a certain person. Consequently, they can by no means be lumped with co-ordinate added entries either.

The main reason, however, why this class of added entries cannot be classified either as co-ordinate or as accessory added entries is the following. The objects proper of the added entries belonging to both latter classes are either books or parts of books; they are in any case documents, that is to say, two-faced objects consisting of a physical and an intellectual constituent part. The objects of the co-ordinate added entries are also physically separate, detached objects, while those of the accessory added entries are, though not separate, yet clearly delimitable within a bigger unit physically, too. The object proper of an added entry belonging to the third class is, on the other hand, no document; it is by no means the translated or the edited book itself, but only one of its aspects, its characteristics, its distinguishing features. It lacks the physical constituent which is indispensable in everything what librarians' terminology subsumes under the word 'document'.

For all these reasons it is impossible not to recognize this third category of added entries, as well as the corresponding third category of elemental objects, as separate and distinct categories. True, this third category of added entries, as well as that of their objects, is, as a whole, considerably less significant than either of the first two; but it is still remarkable enough, especially owing to the circumstance that these objects have no separate material constituent of their own, and constitute thus a class of phenomena of author-title cataloguing the like of which we did not come across up to now. We shall call the objects belonging to this third class 'tertiary elemental objects', and the entries relating to such objects 'tertiary added entries'.

The objects of this third category share with the objects of the first two categories one characteristic: in their capacity as elemental objects they, too, must by all means be retrievable in the catalogue also by their own physical marks. For they are, although lacking a material side of their own, none the less equipped with a physical mark: with the name by which the primary elemental object they are connected with, in visible, physical form designates the person or body involved.

Here we must insert a qualificatory remark. The criterion of the co-ordinate added entry (its approximate equivalence to the pertinent main entry in promoting the retrieval of a primary elemental object) is of a quantitative character, and of a rather vague one at that. It frequently happens that this criterion fails to fix with sufficient exactness the critical point at which the equivalence may be considered as coming to an end. The boundary line delimiting the area of co-ordinate added entries is, however, at the same time also the line of division between co-ordinate and tertiary added entries, and so a vagueness of the former implies automatically a vagueness of the latter, too. What is more, the vagueness of the border line between these two kinds of entries also entails a blurring of the line of division between primary and tertiary elemental objects.

This compels us to raise the question whether our presenting of what we call tertiary elemental objects as a separate, third class of elemental objects is not a mistake altogether? If we refrained from establishing this class, our classification of the objects of cataloguing would become pleasingly simple: all elemental objects would uniformly be objects with a material constituent of their own, and subsumable, thus, under the concept 'document'. But our tertiary elemental objects would have to be classified, then, either as primary, or as secondary elemental objects; and we have already shown that it would not be warranted to take either of these courses. So, I think, we must insist on our proposal of inserting into our classificatory system a third class of elemental objects as well as of entries; simply on the ground that this is a construction which can do justice to the complexities of reality in a higher degree than the opposite one could. After all, every classification and systematization of the facts of reality is bound to be partly unsatisfactory, but one cannot give up looking for the most satisfactory one among them, as it is not possible to attain a lucid representation of reality without making use of classificatory systems.

Besides, the marginal instances in which the status of a particular object or entry is in fact obscure, are comparatively rare, and do not alter the fact that there is a fundamental difference between the respective overwhelming majorities of these two kinds of objects and entries alike.

We do not wish, however, to retard the progress of our course of thought by further pursuing this question of detail; we shall enlarge upon it to a somewhat greater extent in Appendix II, devoted to a more penetrating classification of entries.

44. CAN THE EXISTENCE OF SECONDARY AND TERTIARY ELEMENTAL OBJECTS BE RECONCILED WITH THE CURRENT DEFINITION OF THE FIRST FUNCTION? Since there are no elemental objects – as this has been pointed out earlier, and will presently be corroborated with special regard to the secondary and tertiary elemental objects – which can be classified otherwise than as objects belonging to the domain of the first function; and since, on the other hand, the secondary and tertiary elemental objects cannot by any means be subsumed under that concept of the objects of the first function which the current definition of this function implies – i. e., under the concept 'book', for which we may already substitute the concept 'primary elemental object' – we must now examine the question whether that definition is really as gravely defective as one seems to be obliged to infer from a confrontation of these two facts.

For a first step, we propose to examine a seemingly plausible argument which, if valid, may save and maintain the habitual formulation of that definition.

The services actually performed by accessory and tertiary added entries do not end with the conveying of information about their objects proper, the secondary and tertiary elemental objects; the entries of these two classes enhance the efficiency of the catalogue also in two further ways. First, they procure – as this was already hinted at above, and will be looked at more closely in the course of a discussion of the incidental objects of entries in Appendix II – a more or less valuable additional, indirect access to the respective primary elemental objects. Second, by introducing new elemental objects into the catalogue, they enlarge the bulk of the basic material out of which the composite objects of the second and third functions of the catalogue are built up. These two functions can, evidently, be performed more completely and efficiently, if the catalogue is not confined to registering only the separate editions of, say, a particular play, but also records the editions contained in collected editions of the works of the author, or in anthologies; and if it complements the information about an author's 'oeuvre' by the records of minor, not independently published works, as well as of the translatorial and editorial achievements of this author. This second additional service of the accessory and tertiary added entries is, indeed, of great value. It is not at all impossible that on the whole it is more valuable than even their principal service: the communication of information about their own objects proper.

Now, it might be maintained, and not even without a semblance of plausibility, that the *only* purpose for which accessory and tertiary added entries are prepared is to obtain the first, or the second, or both of these two services which we have represented here as merely additional; and that these entries are not prepared for communicating information about what we have called secondary and tertiary elemental objects, these being no objects of cataloguing in their own right, no elemental objects. If valid, this argument would show a way out of the difficulty presented by the current formula of the definition of the first function which we are here concerned with.

How can it be ascertained whether this argument is or is not valid? Surely not on a historical basis, by an attempt at clarifying the intentions of our predecessors who first applied accessory and tertiary added entries, nor by inquiring into the motives of present-day cataloguers who continue to apply them. It is impossible to ascertain the true nature of the former; while the latter are likely to vary from individual to individual. An attempt at a verification of the argument examined can only be based, therefore, on the services these two classes of added entries in fact perform. On this basis, however, it is impossible not to reject it. As their headings clearly show, the objects which the accessory and tertiary added entries primarily convey information about – the secondary and tertiary elemental objects – are elemental objects in their own right, manifestly different from the primary elemental objects to which they are attached. Another evidence of the erroneousness of the analysed argument is implied in the fact that our catalogues record secondary and tertiary elemental objects, not only under uniform marks, but also under their own physical ones. Codes implicitly insist on that, without conceding any exceptions. Their consideration for the user acquainted only with the physical marks of elemental objects, is in respect to secondary or tertiary elemental objects just as unfailing as in respect to primary ones. This shows that to the mind of codemakers the former two classes of objects are by no means mere components of composite objects, but also elemental objects in their own right. So it is not only in theory that they cannot be classified otherwise, they are treated as such practically, too.

The tentatively raised argument does not serve, then, its purpose; it cannot resolve the contradiction between the undeniable existence of secondary and tertiary elemental objects and the wording of the current definition of the first function.

45. SECONDARY AND TERTIARY ELEMENTAL OBJECTS BELONG TO THE AREA OF OPTIONAL EXTENSIONS OF THE FIRST FUNCTION PROPER. As a next step, let us consider another expedient which might also seem to be able to eliminate that contradiction: an extension of the current definition by replacing in it the word 'book' by our term 'elemental object'.

This expedient will not do, however, either. The latter term not being in current use, this alteration would render the definition practically meaningless, unless complemented by a long specification, an item per item enumeration of all those kinds of primary, secondary and tertiary elemental objects about which information is, or ought to be, conveyed by author-title catalogues. So the task of defining the first function would come to having to include answers to all the questions which we were dealing with in the preceding lengthy survey of the elemental objects. In a theoretical inquiry this would be no impracticable way of proceeding. But what about codes and textbooks?

This is, however, only a consideration of expediency, and so not really decisive. The conclusive argument against the solution in question is that it would substitute for the inadequate old definition of the first function an in a new way inadequate, but substantially still untenable one.

Secondary and tertiary elemental objects differ from primary ones fundamentally, in that the latter must be recorded in any case, on the whole in every kind of library and in all the general author-title catalogues of a particular library, whereas the recording of the former is generally considered not only less indispensable, but in certain kinds of libraries and author-title catalogues directly superfluous. The rules providing for the registration of such objects are, accordingly, mostly optional. The uncommonly wide discretionary powers which are in this respect usually granted to cataloguers, are warranted partly by the considerable differences between the objectives of the different types of libraries and of the different author-title catalogues within the same large library; and partly by the greatly varying importance of the individual specimens constituting most species of secondary or tertiary objects. For instance, a commentary attached to the work which forms the nucleus of the contents of a book, may sometimes be so important that it would be a mistake not to record it autonomously even in the catalogue of a medium-sized public library, another time so insignificant that its neglect will appear quite admissible even in a special research library. This applies to just about all kinds of secondary and tertiary elemental objects.

But if there is such a fundamental difference in importance and role between the primary elemental objects on the one hand, and the secondary and tertiary ones on the other, then it would be an obvious mistake in the definition of the first function to subsume the latter under the same category as the former. We must be careful not only in regard to the functions, but also to the objects of cataloguing, not to mix up things which belong, in consequence of their greatly differing importance, to classes definitely different also in kind.

From this we may infer that the contradiction between the current definition of the first function and the existence of secondary and tertiary elemental objects is only a partial one, and that this definition is, after all, partly correct in representing the 'books', more precisely, the primary elemental objects, as the only class of objects of the first function. They in fact constitute its only class of *indispensable* objects, a recording of which is obligatory under any circumstances. The conveying of information about secondary and tertiary elemental objects the cataloguer may often partially, and sometimes even entirely be dispensed from. The situation might be characterized best by saying that the whole domain of the first function is split up into two sectors of different importance. Its main sector consists in conveying information in any case and unconditionally about all primary elemental objects. Extension and limits of this sector are determined simply and definitively by the range of primary elemental objects, and so the basic definition of the first function, as interpreted in the foregoing chapter, perfectly fits this main sector which might also be called the 'first function proper', or 'the first function in a narrower sense'. This admission, however, we must by no means omit to supplement by emphasizing that the catalogue may, and most catalogues actually do, optionally also fulfil the service of recording elemental objects of a minor rank. Theoretically this service is an extension and a rounding-off of the scope of the first function, though in practice, in individual catalogues, it may be partly or entirely neglected.

Although the service in question is optional only, and not an integral part of the first function, there are two facts which clearly show that it is, nevertheless, a part, a section of the latter. The first of these facts is that the objects in this complemental sector have their foremost characteristic in common with the objects of the first function proper, they, too, are elemental objects. The second, that the method applied in performing this complementary service – the securing of an access to all its objects by their own physical marks as well as by the corresponding uniform marks – is identical with the method employed in performing the function proper.

46. Practical conclusions concerning the codification for the treatment of secondary and tertiary elemental objects. As we have already mentioned, the pertinent rules of most codes are, in accord with generally adopted cataloguing practice, elastic; they contain chiefly optional rules. Quite rightly; in this special respect flexible rules and practice are entirely warranted and must not be regarded as violations of the fundamental principle that cataloguing should be rendered consistent and uniform by cogent rules. Just on the contrary, one might object to meeting in not a few codes compulsory rules prescribing the preparation of certain kinds of accessory or tertiary added entries, e.g., a rule providing for the analytical cataloguing of the works contained in a collection if their number does not exceed three, or of appendices mentioned on the title-page, etc. Such rules draw the boundaries of the recording of non-primary elemental objects of cataloguing much too rigidly and mechanically: on the one hand, they compel the cataloguer to record nonentities, and on the other, they may lure him into feeling absolved from the duty to ponder whether he had not better record, in view of its significance, also a fourth work, or an appendix not mentioned on the title-page. Besides, they disregard the differences between the respective objectives of libraries belonging to diverse types, as well as of the different specimens of the author-title catalogue in a big library. In their ultimate effect they impair the credit and authority of codes, and are thus partly responsible for notions like L. Carnovsky has voiced.

The only maxim which may in this respect have some claim to a relative validity is that the recording of secondary and tertiary objects is of more use in large and research libraries than in small and public ones: this applies equally to each of the above specified three services of the autonomous recording of such objects. Being compiled principally with a view to the needs of large research libraries, many national codes show an in itself not entirely unjustified inclination to give compulsory rules for the application of certain kinds of accessory and tertiary added entries. Since, however, these codes frequently underlie cataloguing in minor libraries too, such rules of theirs tend to result in an altogether unwarranted liberal application of these two sorts of added entries. Another factor likely to have a similar effect is, paradoxically, the effort by all possible means to increase the efficiency of the tools of information. Code-makers must take care that this effort should not mislead them into adopting rigid compulsory rules of a questionable sort.

47. THE RULES FOR PREPARING CO-ORDINATE ADDED ENTRIES UNDER CORPORATE BODIES AND TITLES. Although this topic is no logically integral part of the first objective, of the subject-matter proper of this study, yet I think I may be pardoned for inserting here, on behalf of the second of its objectives, a digression on some rather significant tasks which in the codification concerning co-ordinate added entries have so far not obtained satisfactory solutions yet.

The rules of our codes concerning this class of added entries are, in general, satisfactory and explicit enough. There is, however, an important exception. By displaying a lot of grave mistakes and defects, the rules concerning *corporate* co-ordinate added entries differ in almost all codes just as strikingly and unfavourably from the corresponding provisions of their other chapters, as most other rules of their chapter on corporate authorship do. The first code to which this does not apply are the new *Anglo-American Cataloguing Rules*.

One of the most conspicuous mistakes habitually made in this context is the omission of an imperative rule providing cogently for the preparation of an added title entry if the main entry has a corporate heading. What is more, regarding some particular cases many codes even explicitly prohibit the preparation of such an added entry. It is, I think, clear that, if a code-maker does not feel capable of abstaining from the use of corporate main entries, he must at least not forget that these are never able to reach the efficacy of personal author main entries, and that it is, in consequence, not permissible to desist from providing the supplementary second handle which a title added entry may furnish, the efficiency of which on the average equals the efficiency of the corporate main entry. This is a maxim no less obvious than the reversed one, that if a corporate publication proper is entered under title, it is inadmissible omitting to make a corporate added entry. True, there may be cases in which the utility of the here advocated corrective to the corporate main entry is practically negligible; these cases are, however, comparatively rare, and therefore – in view of the cardinal principle that exceptions to rules must be conceded only very sparingly and for very weighty reasons – it is not warranted on their account to make any exceptions from the general rule.

Perhaps less conspicuous, but without doubt no less harmful than the former is another shortcoming shared by many codes, its effects being felt in practice even more frequently. This shortcoming consists in the lack of adequate provisions for the treatment of the extremely frequent cases of the participation of several bodies in bringing a publication into existence. In our days cataloguers have very often to deal with publications which are a result of co-operation between three or four or even more corporate bodies that have all played some role of intellectual nature in their production. But – excepting the *Anglo-American Cataloguing Rules* – there is no code that would advise them what to do in such a case. It is hardly necessary to emphasize that this is a matter very far from being insignificant – not only the satisfactory recording of primary elemental objects, i.e., the fulfilment of the first function, is at stake here, but also the adequacy of the fulfilment of the third function.

With the many open questions of corporate authorship I have, however, dealt at some length in an earlier paper already.[45] Here I confine myself, therefore, to drawing briefly those inferences from the findings of that paper which have a direct bearing upon the problems of co-ordinate added entries. It follows from my basic finding according to which the corporate *main* entry ought to be jettisoned altogether, that the large libraries which we are concerned with, should make added entries – a considerable part of which will be co-ordinate in character – under all bodies which played a not definitely insignificant role in bringing into existence either the primary elemental object in hand or the work contained in it, if they are named openly (i.e., in the preliminaries, including foreword and introduction, the colophon, or on the cover or binding) by that object; regardless of the nature of their role. Exceptions are to be made with most publishing firms, with bodies whose participation is confined to a pecuniary grant, in case of a pullulation of participating bodies, and finally with most elemental objects which are entered under the personal author of the work they contain. In certain cases, on the other hand, even bodies not named openly should be given corporate added entries (a symposium the proceedings of which the publication contains; bodies involved only in the production of a secondary elemental object).

Among the practically significant problems connected with co-ordinate added entries there is, finally, one which it would seem to me very opportune to reconsider. This is the questionableness of the practice not to complement personal author main entries by title added entries, except when the author is a hidden or a pseudo-author; a practice which, excepting the United States and the countries adopting American cataloguing practice, seems to be prevailing everywhere in the world. It seems to me rather doubtful whether this is – at least in libraries using unit cards – an economy warranted in our days. Apart from the personal author entry, the title entry is indisputably the most efficient device for locating an object in the author-title catalogue. Many catalogue-users whose knowledge of the author's name is not accurate enough, *are* aware of the correct title of the item they seek. A catalogue which follows the present prevailing practice, rather unpardonably fails these users. With the present-day tendency of preparing a great many added entries and references merely or chiefly for the benefit of the users who approach the catalogue with but an imcomplete knowledge of the formal marks of the item they need, this practice is not in harmony at all. A rule compulsorily prescribing the preparation of title added entries even if the main entry is a personal author entry, would be likely to increase the efficiency of the catalogue to a

[45] DOMANOVSZKY, Á.: Die korporative Verfasserschaft. *Libri*. Vol. 11, No. 2, 1961. Cf. also § 67–71 of this study.

greater extent than any other conceivable improvement not involving considerably higher costs might do.[46]

48. THE FUNCTION-STATUS OF THE FIRST FUNCTION. The above analysis of the objects of the first function includes (with regard to this function) already a reply in the affirmative to the first question raised in § 11. It demonstrates conclusively that what cataloguers are used to call the first function is really a function in the above fixed sense of the word.

There are two further cardinal questions concerning the place of the first function among the functions of the catalogue, which we shall have to answer: whether the first function is in itself a separate and distinct function, and not a part only of a function with a more comprehensive scope; and whether it is really warranted to rank it as first among the three functions. It is, however, obvious that we shall not be able to answer these two questions until we have finished examining the other two functions.

[46] Cf. TATE, E. L.: Main Entries and Citations. One test of the revised Cataloging Code. *Library Quarterly*. April 1963. – *The Requirements Study for Future Catalogs* carried out recently at the University of Chicago Graduate Library School produced even more cogent arguments in favour of such an innovation than Mrs. Tate's test did. Two of the studies prepared in the framework of this research project (by Dolores K. Vaughan and Mary Grathwol) have come to the astonishing result that in regard to the average search length involved, titles are very much superior to authors' names; and according to Mrs. Vaughan they were not inferior to them even in memorability. See the summary and overview of the Study, given by Don R. Swanson in *Library Quarterly*, Vol. 42, No. 3, July 1972, pp. 302–315.

CHAPTER VI

THE SECOND FUNCTION: THE RECORDING OF 'WORKS'

49. GENERAL REMARKS CONCERNING BOTH KINDS OF COMPOSITE OBJECTS. Before addressing ourselves to the task of examining the second function, we must make a preliminary remark which applies to both the second and third functions.

We have mentioned introductorily that by taking stock of all the elemental objects (i.e., of all the objects belonging to the proper field of the first function, as well as to its extensions) we get at the same time an answer also to one of the cardinal questions concerning the objects of the second as well as of the third function (i.e., the composite objects). Since the sum of the elemental objects recorded in any given catalogue definitively determines the aggregate of the components at the disposal of this catalogue for building up composite objects (the latter consisting exclusively of elemental objects), a survey of the elemental objects is bound also to yield an answer to the question as to the total extent of the respective fields which the second and the third functions can and should cover.

We must, however, emphasize that this naturally does not imply a one-sided dependence of the actual scope of the second and third functions on factors extrinsic to them. These functions play, of course, a considerable part in determining their own scope. For instance, one of the springs which have brought about the spread of analytical cataloguing (in the narrower sense of the term) was very probably the uneasiness cataloguers must have felt whenever omitting to record an edition of an important work published not separately but in a collection. If this assumption holds true, the resulting extension of the coverage of the second as well as of the third function has been brought about by the intrinsic logic of these functions themselves. Notwithstanding this, as soon as it is entered analytically, i.e. autonomously, every item of this description becomes not only a component of the corresponding composite object(s), but also an elemental object – in spite of its autonomous recording having been brought about, partly or solely, by motives rooted in the second and third functions. The influence exercized by these functions upon the expanse of their own coverage is thus indirect: they directly influence the range of elemental objects, and the extension of this range automatically entails an increase in the sum of the components of the corresponding composite objects, a growth of these objects.

It is no less obvious that, although the aggregate of its elemental objects definitively determines the aggregate of the elements at the disposal of a catalogue for building up the aggregate of its composite objects, it does by no means determine the selection, the range, the assortment, of the components of *any* of the *particular* composite objects recorded in this catalogue. It will be the task of this chapter and the next to investigate the principles governing the composition of the individual composite objects and by their findings to reveal the essence of the second and third functions.

50. Recapitulation of what we have hitherto found out about the second function. Now coming to the second function, let us first restate shortly the pertinent facts we have already become acquainted with in the preceding chapters.

According to the current definition, the second function consists in conveying pooled information about all the editions of any work available in the library, that is to say, in relating all the single units of information about the single editions of a given work so that they form one new, composite unit of information. The object of such a unit of information, the individual object of the second function itself, is, naturally, also of composite nature. In one respect, however, the objects of the second function resemble the objects of the first function. They, too, have two opposite aspects relevant to cataloguing; a material and an intellectual, i.e., immaterial one. Viewed in its material aspect, each individual object of the second function is the sum of a number of elemental objects which are separate or at least delimitable physically, a set of books and portions of books which all contain the same work. In another of its aspects an individual object of the second function is an intellectual unit: the work itself. In contradistinction to the first, with the second function it is this intellectual side of its objects which is really relevant to, and more strongly emphasized by, cataloguing. The users' needs to be satisfied by the second function are focused on this intellectual side; and it is this side which determines the selection of the elemental objects which are to compose a particular object of this function.

The method applied to performing the second function is adapted to the dominantly intellectual nature of its objects. As we have seen, this method consists in disregarding the physical marks of elemental objects and in recording them as components of composite objects consistently under the same set of formal marks – the uniform marks – which have been introduced by bibliographers and cataloguers directly for the purpose of providing a satisfactory access to the intellectual constituent, the content element of these objects. Uniform marks might be regarded, therefore, as formal marks attached to this intellectual constituent, i.e., to the work itself. The element of arbitrariness which undeniably adheres to the use of uniform marks, can be, and in fact is, mitigated by choosing for this role the best known and mostly used ones among the formal marks connected with a given work. These are usually the physical marks that appear on a majority of its editions, but they may also differ from them.

Parenthetically mentioned, for the sake of international uniformity this general rule has to be submitted to a radical qualification: the choice of the uniform marks must exclusively be based on the formal marks of the editions published in the original language of the work. But even this compromise cataloguers cannot stick to consistently and without reserve. By reason of linguistic difficulties and practicabilities, the said restriction imposed upon the general rule must suffer, in its turn, again restrictions in favour of the general rule. For instance, in most catalogues of the countries using the Latin alphabet the editions of oriental works are not collected under their formal marks in the original language, but under those of a translation written in Roman characters.

51. THE REASONS FOR ADOPTING THE SECOND FUNCTION, AND THE DEVICES USED IN PERFORMING IT. The great majority of those users of the catalogue we usually describe as "looking for a particular book", look in fact for a particular text, a particular work.[47] To illustrate this by an extreme example: incomparably more users ask for any undefined edition of *Hamlet* than for the first quarto, or any other particular edition. The majority of the most important class of the users of the author-title catalogue are, then, fully satisfied if they obtain any one of quite a number of particular books bearing not only widely different physical marks, but having frequently even considerably differing contents. But they can only be served adequately by a catalogue which informs them in an appropriate manner about *all* those available editions of a work which, suiting their needs equally, are interchangeable for them. If the catalogue fails in this, it may easily occur that a patron will have to leave unsatisfied, because those editions of the work needed which he did find in the catalogue happened to be out on loan, while of the availability of other editions standing all the while in the stacks he remained unaware, as he could not find their entries. It is in research libraries that an adequate performance of the second function is of the greatest consequence. On the one hand, research workers are served by a translation of a particular scientific work mostly just as well as by an edition of the original; and on the other, they form a class of catalogue-users for whom it is more important than for any others that they should be enabled to select among all available editions that particular one which suits their purposes best – as to its age, as to the commentaries, the critical apparatus, the bibliography included in it, etc.

We have spoken here only about an adequate, and not about a perfect fulfilment of the task under consideration. As usual, the cataloguer must content himself, here too, with optimality, i.e., with attaining a degree of relative adequacy, determined by the necessity to balance the comparative utilities of the diverse assortments of services his catalogue may perform against each other, as well as against the limited means at his disposal. Therefore, he must draw the limit of the fulfilment of the second function at the point where the information contained in the processed elemental objects themselves and in standard works of reference ends. And what is more, his going farther is directly undesirable, seeing that he has to deal – in L. Jolley's words – "not in new facts but in generally accepted data".[48] Let us add, anticipatively, that the same restriction applies to the third function as well.

[47] It is, perhaps, not unreasonable to anticipate the objection that the majority of works are published only in a single edition, and that, since in the case of such works the user *is* looking practically for a particular book, it is not warranted to say, as I do, that the majority of users *is not* looking for a particular book. This objection, however, does not hold, since from the point of view of the problem under consideration such cases are simply beside the point and should be disregarded entirely; but also because the majority of requests are in most large libraries focused on a minority of the works in the holdings, and a considerable part of this minority is composed of frequently or at least repeatedly published works.

[48] Cf. op. cit., p. 10.

The appropriate mode of conveying information about the objects of the second function consists, as we know, in collocating in the catalogue all information about the editions of any particular work; in this context it is indifferent by which of the three forms of entry this task is performed. In a non-collocating catalogue it is very often only the exceptionally expert and pertinacious reader who can succeed in retrieving all the available information about a particular work. Such a catalogue is apt to become misleading, simply because readers are likely to expect a strict collocation of all information on a particular work. A patron finding under a particular heading one or several entries recording editions of a particular work, may easily infer that the library does not possess any other editions, and conclude his search prematurely, with an entirely or partly unsatisfactory result. It would be an unpardonable mistake to raise against this argument the objection that it is the duty of the reference service to prevent the occurrence of this; first, because that service does not intervene between reader and catalogue except in a fraction of cases, and second, because the catalogue must on principle afford him a direct access, without any intermediaries. This principle forbids the use of all practices which may occasionally turn misleading.

As long as the physical marks of all the editions and variants of a work are identical, the fulfilment of the second function does not require any special procedures or additional entries. The entries made under physical marks, i.e., the entries whose principal purpose is to serve the first function, are assembled in such a case automatically; and so the second function is performed incidentally, and remains practically latent. Not so, however, if the physical marks borne by the various editions differ; it is in this very common situation that the need of the application of uniform marks, i.e., of a rather elaborate technique contrary to the one used in fulfilling the first function, arises. We might add parenthetically that sometimes this technique may be employed even when the physical marks of all existing editions are identical. This may occur, if the majority of reference works designate a particular work by formal marks differing from the physical marks consistently used in its editions.

52. THE WORD 'WORK' MUST BE QUALIFIED, BECAUSE UNQUALIFIED IT IMPLIES AN UNDERLYING CREATIVE ACTIVITY. We come now to the topic of the scope of the second function.

Since we have chosen for our point of departure the current definition, according to which the second function consists in conveying pooled information about all the available editions of a work, our next task must, obviously, consist in clarifying the meaning of the word 'work' and of the expression 'all editions of a work', both contained in this definition. By clarifying the first we shall attain a confirmation of the answer reached above to the question as to the boundaries of the total coverage of this function, as to all the classes of objects relevant to it. By the clarification of the second we shall disclose the principles which govern the composition of its single objects, the selection of the elements of which these objects have to be built up.

If we compare the current definitions of the first and second functions, we find that the word 'work' in the latter can denote only the intellectual side, the content element of the phenomenon designated as 'book' in the former. The word 'book' denotes, as we have seen, in everyday language as well as in the terminology of cataloguing, a phenomenon in which a material and an intellectual side are necessarily and undetachably connected. The word 'work', on the other hand, is used in everyday language (along with a dozen here irrelevant ones) in two distinctly different to our subject relevant senses. The first is exactly congruous with the everyday meaning of the word 'book'. Any adequate and expedient cataloguing terminology ought, obviously, to abstain from using the word 'work' in this sense, as a designation for a thing for which it has already another, much more appropriate term. When used in the second sense, the word denotes only the intellectual constituent, the content element, of the two-sided phenomenon 'book', conceived as a thing detached from and independent of the material vehicles with which by the word 'book' it is represented as undetachably united. In this second of its senses the word 'work' designates, thus, an abstraction, an immaterial, intellectual reality, which is, however, catalogically extremely relevant. Its use with this particular denotation is warranted by the fact that, although a 'work' is in reality always merged with a material vehicle, it is never confined to any *particular* one, and may materialize at the same time in several different such vehicles. Its existence is independent of each of its materializations, of every one of the books which contain it. In a catalogical sense, books are unique non-recurrent material realities; their physical form is fixed once for all, they are quite incapable of changing. If a book is reprinted from the original setting of type with only the date of printing altered, for the cataloguer and bibliographer it becomes already a new book, no more identical with the original one. A 'work' in the second of the above senses of the word, on the other hand, may change freely – it may be revised, enlarged, abridged; its formal marks, its title or the name of its author, may be altered; and even the person(s) of the author(s) may change – without the work loosing its identity, without its becoming a different, a new work.

In the current formula for the second function the word 'work' is used in this latter sense. This is to be inferred from the facts that, first, the use of the word 'book' in the definition of the first function excludes the possibility that in the formula for the second function the word 'work' should be used with a denotation which in the former was borne by the word 'book'; and second, that up to now no attempt has been made to attach to the word 'work' a specific, technical, catalogical meaning different from the above, its alone relevant everyday meaning.

But the denotation the word 'work' bears when it is used in this sense in everyday speech, does not seem to include indiscriminately all documents without regard to the quality of their intellectual components, but merely those which have been brought into existence by some kind of creative mental activity, usually that of an author. In everyday usage 'work' in this second sense, and 'author' or 'creator', are correlated concepts. Although usage draws the boundary line of creativity in this context rather liberally, yet, it presupposes at least a faint trace of intellectual creatorship, so that the contents of some books do not seem to be considered 'works'. From this being so one should infer that the total coverage of the first function must, in this respect again, have necessarily a greater expanse than that of the second, seeing that the former includes all books, while the latter excludes a particular class: the books the contents of which are not usually called and considered 'works'.

To draw the dividing line sharply between activities that are creative and those that are not, is, naturally, impossible. A novel presupposes creative activity, so it is indisputably a 'work' in the everyday sense. To a single issue of a literary review the designation 'work' is never applied, though it contains a number of unquestionable works. Viewed from the side of these two cases only, the everyday meaning of the word seems clear and indisputable. In many instances, however, opinions on the question whether or not one can call the contents of a particular publication a 'work', are liable to differ; take, e. g., a telephone directory, a time-table, a college almanac, etc. In the case of certain kinds of publications closely related to each other, the transition from doubtlessness to dubiousness of their status as 'works' is almost imperceptible. To a good anthology of modern verse one applies this designation unhesitatingly; if confronted with a charitable publication for the benefit of the flood-damaged, which contains verses by contemporary poets who happened to be willing to contribute, one may already hesitate; but most people will decline to apply the designation 'work' in this second of its senses to an almanac edited by way of advertisement by a firm of publishers and containing a selection from volumes of verses published by that firm. And the gaps between these three still clearly distinguishable types may easily be filled in with further intermediate forms produced by the modern book trade in great variety. The task of determining the exact spot where such a continuous chain of gradated forms, every one of which differs from its neighbours only very slightly, ought to be split up into two parts, in our case into 'works' and 'non-works', is a very precarious one: it can be solved only arbitrarily, and is in consequence apt to be solved in a great many diverse ways.

53. UNDERLYING CREATIVE ACTIVITY IS NOT A NECESSARY CHARACTERISTIC OF THE OBJECTS OF THE SECOND FUNCTION. On the basis of the second everyday meaning of the word 'work' it is, then, impossible to determine the exact scope and limits of the second function. This impossibility does not entail, however, any unfavourable practical consequences, since, in performing the second function, cataloguers, as a matter of fact, do not wish to distinguish between 'works' and 'non-works' in the above sense. The creative character of the exertion which has brought the contents of a book into existence is irrelevant to the selection of the components of the individual objects, and consequently also to the determination of the general limits, of this function. Current cataloguing practice treats publications which have their place at the 'creative work'-end of the scale in exactly the same manner as those placed at the opposite end; every document which is recorded in the catalogue autonomously, as an elemental object, is also automatically subjected to the treatment which the second function calls for. That Aldous Huxley's *Point Counter Point* should be considered a work, and so an object of the second function, is quite in accord with the everyday meaning of the word 'work'. On the other hand, a publication bearing the title *Yearbook for Pedagogics* and containing, say, ten papers devoted to topics taken from sundry distant sectors of the domain of pedagogics, absolutely unrelated to each other, seems to be no 'work' in the second everyday sense of the word, but only a publication, a book, containing ten independent works. In spite of this, modern cataloguing practice is not content with entering it and its reprint published under the new title *Studies in Pedagogics*,

according to their respective titles in two different places of the catalogue, but insists on their being related and brought together, just like the diverse translations of *Point Counter Point.* And it cannot be denied that this practice is absolutely sound, practically as well as logically. If one acknowledges, very rightly, the obligation to convey pooled information about the editions of a novel, it would be quite unwarrantable to repudiate it with respect to the pedagogical collection in question.

54. WORKS PUBLISHED IN COLLECTIONS DO NOT USUALLY BECOME CONSTITUENT PARTS OF THE OBJECTS OF THE SECOND FUNCTION. In the two preceding paragraphs it has been shown that the current definition of the second function is inadequate, since it determines the total range of the objects of this function not only nebulously, but in one respect also inaccurately. Its failure to give a correct answer to our second cardinal question, i. e., to fix the principles governing the composition of the single individual objects of this function, is, however, even more glaring. Apart from being in this respect too nebulous and partly inaccurate, this answer is with regard to a rather important point also directly misleading.

We must refer here once again to the fact that cataloguing practice generally neglects to register most of those editions of intellectually self-contained, separate works which have not been published in a physically separate form; that it refuses to render them autonomous objects of cataloguing. For the second function this practice has a particularly striking consequence: it occurs very often that a catalogue does not record but a part of the available editions of a work, and abstains from providing access to another part, which would be of no less use to patrons. Among the hundreds of entries of Thomas Mann's works in the catalogue of a large library there are usually only one or two secondarily filed under the title *Tonio Kröger,* those which record editions published in the form of separate booklets or in collections containing not more than two or three stories. A large number of other editions of this story, incorporated in collections either of Thomas Mann's works, or of German short stories, or else of the best short stories of the world, are by most catalogues suppressed entirely or partly. Entirely, if the collection is an anthology the entry for which is not headed by the name of Thomas Mann; partly, if it is a collection of works by Thomas Mann, but among the several hundred entries headed by his name there is no added entry sub-filed under the title *Tonio Kröger.* In the latter case to patrons who are tenacious enough, the name of the author still provides a clue, though not a very efficient one. This practice is in regard to anonymous works, which it makes absolutely irretrievable if they are not published separately, much more inadequate than in regard to authors' works.

Moreover, this practice completely suppresses the works which have been published exclusively in collections, and never separately. As regards such works, our catalogues are not so flagrantly inconsistent as in the case of *Tonio Kröger;* but the contradiction between the wording of the current definition of the second function and actual cataloguing practice becomes here even more glaring. Items which quite indisputably have to be classed as 'works' are left in such cases *entirely* unrecorded.

I think we have already made it clear enough that the current definition of the second function, just like that of the first function, though hitting the target approximately, is yet very far from being correct or adequate. Instead of determining with the greatest attainable accuracy the scope and limits of the second function, this definition, too, stops short at indicating merely the function's most conspicuous aspect, its core. In short, instead of defining it, it only labels it rather vaguely.

We might mention in passing, that this is an attitude one rather frequently meets with in cataloguing theory as well as in cataloguing codes. Authors in both fields are inclined to be content with taking only the most typical constellations and forms within the compass of a particular problem into consideration, and not to bother about examining its total compass, inclusively of the less typical forms and constellations occurring in its peripheries. In consequence they often fail to express their real intentions with the requisite clarity, and abandon, thus, unnecessarily broad border-zones to arbitrary decisions of the catalogue-worker. It is evident that such an attitude is bound to render the resulting practice inconsistent to an extent which could be diminished considerably.

55. Excursus on the use of the word 'work' in present-day cataloguing terminology. The definition of the second function is already the second instance in which we come across an incorrect use of this word (and we might mention anticipatively that in our analysis of the third function we also shall have to deal with a third instance). The first instance was its use, instead of the word 'book', as a designation for the primary objects of the first function. But this particular terminological laxity makes its appearance inconsistently; whereas the incorrect use of the word in the definition of the second function together with the resulting inaccurate definition, is adopted in principle and consistently maintained. Throughout the world authors keep using the word 'work' instead of a special technical term the meaning of which would have, in two diverse respects, sharply to deviate from the former's alone relevant everyday meaning. And they do that without ever troubling to say so; just as if they were using it in the latter, in the everyday sense. The resulting discrepancy between their definition of the second function and their actual practice did not disturb them sufficiently to induce them either to drop the word 'work', and to replace it in the definition by a special technical term, or else to specify at least explicitly that particular meaning, different from the everyday one, which they in fact mean to give to the word in this context.

This will, however, astonish only those who are not aware of the fact that also the most prominent codes and many contemporary authors are lagging far behind even the rather modest terminological tidiness represented by the current formula. As a matter of fact, most codes and authors insist on using the word 'work' simultaneously in three widely differing senses: to denote, first, that two-sided thing composed of a material and an intellectual constituent which ought to be called always and consistently a 'book'; secondly, the detached intellectual constituent part only of this compound thing; and thirdly, both these conspicuously different things, the whole and one of its constituent parts, at the same time.

As a striking example of this untidy terminology one might mention the *Prussian Instructions* which, in spite of their being undoubtedly one of the most outstanding achievements in the history of cataloguing codification, unflinchingly and steadfastly keep employing the word *Schrift* alternatively with the three different meanings just mentioned. The *Prussian Instructions* were, however, formulated seven decades ago, and it might seem natural to suppose that in the meantime all the gross negligences of this sort have already completely been eliminated from our professional terminology.

As a matter of fact, they have not. Cataloguers continue to stick to a terminology the inadequacy of which has already been demonstrated decades ago. Let us illustrate this curious fact by the new *Anglo-American Cataloguing Rules.*

The *AACR* are based, avowedly, on Lubetzky's ideas and on the *Paris Principles.* In the latter the definitions of the functions of the catalogue are founded on a sharp distinction between 'books' and 'works', the respective objects of the first and the second function.[49] In spite of this, the *AACR* stick firmly to the old custom of using the word 'work' alternatively with the above specified three diverse meanings, without the slightest trace of consistency, and naturally without ever specifying that particular meaning with which it is used in the given case.

To designate the thing the *Paris Principles* call a 'book', the *AACR* use mostly the word 'work'; on other occasions, however, the expressions 'edition of a work' or 'publication'. In the Introductory Notes to Chapter I of *AACR* (p. 9.) you find the passage: "The entry for a work is normally based on the title-page, or any other part of the work". Here the word 'work' is, obviously, used to denote the 'book' of the *Paris Principles*, seeing that a 'work' in the Paris sense cannot have a title-page. In §1A, which is devoted to the most fundamental rule of the whole code, the word 'work' is used within two and a half lines four times, each time inaccurately, instead of the word 'book'. This is clearly shown by the fact that in the first three instances it is the object taken by the verb 'enter'; although a 'work' in the catalogically exclusively possible sense of the word cannot be entered, only a 'book' can; while the fourth time it occurs in the phrase "the author, whether named in the work or not", which makes it even clearer that it is used instead of 'book'. In the next sentence, in subsection B of the same paragraph, it is no more the word 'work', but the word 'publication' which is used to denote the same thing. In §2A 'work' denotes undoubtedly 'work' in the Paris sense; and the 'books' of the *Paris Principles* are clearly distinguished from it by being called 'editions of a work'. In the preliminary note to §3 'work' signifies, in my opinion, again 'work', but one might as well maintain that its denotation includes both 'work' and 'book'. In subsection B2 of the same paragraph and in §4A the word is used without any doubt instead of 'book', seeing that these paragraphs keep speaking of "the title-page of the work".

The vacillation between three diverse meanings of the word 'work' which we have shown to be exhibited by a mere four initial paragraphs, continues throughout the whole Code. In Chapter II the often reiterated phrase "the form of name appearing in the works of an author" shows that here the word stands for 'books'; whereas according to Chapter III the forms of the name of a body are to be found in the 'publications' of this body. In Chapter VI – which is devoted to the descrip-

[49] *Statement of Principles*, p. 91 (Cf. our Note 13). True, in its subsequent paragraphs the SP did not stick constantly enough to this terminological distinction which it had made in § 2 very clearly; but in the meantime this inconsistency has already been put right by its Annotated Edition.

tion of 'monographs' (this is again an instance of a rather questionable employment of a term the firmly established everyday meaning of which is quite different!), and cannot be concerned, therefore, with 'works' (since these cannot be 'described'), but only with 'books' – the word 'work' is *almost* consistently used instead of the terms 'book', 'publication', or 'edition of a work'. Not *quite* consistently, though, as this was only to be expected. "To distinguish one edition from another of the same work" – according to the introductory notes to this chapter – "one or more of the following must be known:... some detail of physical description, such as the number of pages or volumes in the work". Here again the word 'work' is used – in a single sentence! – in two distinctly different senses.

The testimony of *AACR* furnishes, I think, convincing evidence of the fact that a big majority of cataloguers, including really eminent ones, still continue habitually to overlook the axiom that it is impossible to think correctly until one has learnt to call things by names which have a definite and precise meaning. This implies not only that a particular word should denote always the same thing (or species or genus of things), but also that one had better denominate a particular thing always by the same word.

I am afraid, many cataloguers might consider my holding that axiom in so great veneration as well as my laying so great a stress on its application to the use of the word 'work', as unjustified and quite superfluously pedantic. They might argue that in spite of their having taken no notice of the axiom, and used the word with an equivocal meaning, they have managed quite well to get on, for centuries by now. So, I think, I had better illustrate by a few examples the unshakable validity of the axiom, and at the same time also the fact that my insistence on the impropriety of using the word 'work' ambiguously, cannot by any means be assessed as mere pedantry, since such a use is bound to entail rather unpleasant practical consequences.

Let us take first a school-book one half of which contains excerpts from Livy selected by a certain Mr. Brown, the other a lengthy introduction, devoted to Roman historiography and to the biography of Livy, and copious notes to the excerpts – all from the pen of that Mr. Brown. Now, if you use the word 'work' ambiguously, the general rule providing for a 'work' being entered under its author, has, as regards the textbook in our example, two entirely different meanings. If you attach to the word 'work' the denotation which alone should be attached to it in cataloguing terminology, you must conclude that our textbook has two authors: Livy and Brown, since works by both are contained in it. So you have to apply the rule providing for entry under the principal author, and to enter it under Livy. Not so if you happen to allot to the word 'work' its alternative meaning: the meaning 'book'. In that case you must regard Brown as principally responsible for the existence of that textbook, which you must enter under his name.

Another example. Let us consider a rule with the following wording: "If the majority of the works of an author have been published under one pseudonym, this should be chosen for his uniform name". If you hold on to the equivocal use of the word 'work', this rule, too, becomes necessarily ambiguous. If to the word 'work' its catalogically proper meaning is assigned, that rule means that the pseudonym of an author must be chosen for his uniform name if only a majority of the *first editions* of his works have been published under it, even if these are by far outnumbered by later editions published under his original name (which, of course, often happens with classics frequently published in recent times). If, on the other hand, the word 'work' in the rule is meant to mean 'book', the very same rule calls for just the opposite choice of that author's uniform name.

Another example is provided by rule 4 of *AACR* on the treatment of 'works published under editorial direction'. In the wording of this rule the word 'work' is openly used alternatively in two different senses; and this leads, of course, to the meaning of the rule becoming occasionally obscured. Let us take, for instance, a work of multiple authorship the editor of the original edition of which is unknown, while its second edition has been published on the initiative of, but only slightly revised by, a new editor (who designated himself as its editor on the title-page). If the rule is meant to refer to 'books', that second edition is to be entered under its editor; while if it refers to 'works', the same book must be entered under title.

We might conclude this digression by trying to offer an explanation for the very astonishing fact that even by the really glaring contradiction between the definition of the second function and the actual treatment of works not separately published, cataloguers have never been brought to perceive the necessity of replacing the word 'work' in that definition. The most plausible explanation may be found, in my opinion, in another of the ambiguities of the meaning of this word, which our professional terminology has not cared to eliminate either (e. g., by an employment of qualificatory attributes, for instance: 'primary' and 'secondary' works): in our calling the story *Tonio Kröger*, and the anthology in which it has been published, indiscriminately, a 'work'. It may have been this linguistic fact which has diverted the attention of codifiers and theorists from a significant catalogical fact. Having recorded an edition of an anthology not only under its own physical marks, but, if necessary, also under its uniform marks, they might have got the impression that, so far as this particular item was concerned, they had adequately attended not only to the first, but also to the second function, to the function of recording 'works', and eased by this impression, they might have failed to perceive that, by suppressing the separate, self-contained works contained in the anthology, and by tacitly discriminating between two diverse kinds of works, they have generated, or at least acquiesced in, a practice which was quite incompatible with their definition of the second function. This applies, naturally, not only to the short stories published in anthologies, but to every intellectually self-contained work published as part of a collection, e. g., of a historical *monumenta*, of a festschrift, of the proceedings of a symposion, etc.

56. A POSITIVE ATTEMPT AT CLOSING THE GAPS OF THE CURRENT DEFINITION OF THE SECOND FUNCTION. PRELIMINARY REMARKS. We shall now try to fill the gaps of the current formula and to determine the scope and limits of the second function with every possible accuracy.

It should be mentioned preliminarily that for the word 'work' some authors have in this context substituted the term 'literary unit'. In itself this was an ingenious move, because a new term, unknown to everyday language, could have been given any clear-cut meaning which the terminological needs called for. It is, however, just at this point that these authors failed; they omitted to give their new term a clearly outlined meaning. So they were, of course, bound to fail in solving the task in hand: in determining the scope and limits of the second function with the necessary accuracy.

Another preliminary remark. It was already mentioned that the second function is concerned first of all with the contents of documents, irrespective of their being or not being a result of creative activity. Would it not be possible, then, to amend the current definition by simply substituting in it for the word 'work' the word 'contents' or 'text' – by saying that the second function consisted in conveying assembled information about all editions of a given text, or more generally, about all the documents which have the same contents? Since every document has some contents and is in principle a component of one of the objects of the second function, by altering the wording of the definition in this manner it would seem possible to overcome effortlessly all the difficulties of answering the question as to the aggregate of the objects, and together with this also the question concerning the boundaries of the general scope of this function. According to a definition altered in this way, these boundaries would be congruent with the boundaries of the first function enclosing the aggregate of elemental objects; and so the question under consideration would become identical with the question regarding these latter boundaries, which, however, has already been answered by our survey of the elemental objects.

This is in fact so; but the suggested solution is nevertheless impracticable. Altered in this manner, the definition would imply that the second function were confined to conveying pooled information about the different materializations, the different editions of the very same text, of entirely identical contents, only. It would apply to books with different formal marks but identical contents, e. g., to two books containing nothing else but the original text of *The Pickwick Papers*, published under the pseudonym Boz and the name Dickens, or else the bare text of Voltaire's epic, published under the titles *La Ligue* and *Henriade*, respectively. Current cataloguing practice is, however, not satisfied with relating and displaying together diverse elemental objects with entirely identical contents only. Most cataloguers deem it indispensable for the entries of all available editions of *The Pickwick Papers* to be collocated in the catalogue, regardless of their language as well as of the accessories attached to the original text (introductory study, commentary, philological notes, etc.). They want to enter the first edition of a work in the same way as its enlarged new edition, even when there appears in the latter a new co-author who has undertaken the task of revising the original work. Or again, the contents of the *Oxford Book of American Verse* have been altered in a new edition, and also the person of the compiler has changed, nevertheless, cataloguers do not consider this new edition a work different from the first edition, and insist, accordingly, on treating both as components of the same particular object of the second function.

These illustrations show that the elemental objects to be brought together by the second function must be connected with one another by the identity of a nucleus of their contents; which necessarily implies that they must have in common, at least partly, also the intellectual source of their contents. But they demonstrate also that the limits within which the character as well as the closeness of both these kinds of relationships may vary, lie very wide apart. The relationship constituted by the common intellectual nucleus of their respective contents may vary, for instance, between a complete identity of these contents and an absolute lack of any *literal* identity.

As a result of this intricate state of things, the suggested replacement in the current definition of the word 'work' by the word 'contents' or 'text' is, obviously, not feasible, because it would explicitly – and quite wrongly – exclude all variants mentioned in our examples, their contents being not entirely identical with the contents of the respective original editions. So it would, indeed, eliminate one of the shortcomings of the current definition, and render its answer to the *first* cardinal question concerning the scope of the second function unequivocal. But it would at the same time replace this shortcoming by an even graver new one. Altered in this manner, the definition would give to the *second* cardinal question concerning the scope of this function, i. e., to the question as to the composition of its individual objects, an answer entirely incongruous with the real intentions of cataloguers and with general practice.

57. AN INTERMEDIATE STEP IN OUR ATTEMPT TO CLOSE THE GAPS OF THE DEFINITION OF THE SECOND FUNCTION. We shall now make our positive attempt to complete the current definition, so that it should determine the scope and limits of the second function with the utmost possible accuracy.

Our foregoing results enable us to attack this task from a rather advantageous position. We must not worry any more about the extent, the range, of the raw material available for building up the *sum* of the objects of the second function. This question, which the current definition has not been capable of solving adequately, is already answered by our finding that the aggregate of all the components of the objects of the second function is in every catalogue determined by the aggregate of the elemental objects recorded. This means that the second function can embrace only such components of its composite objects which appear in the catalogue also in the capacity of elemental objects; and that, on the other hand, it *does* embrace these in any case, entirely irrespective of the circumstance whether the activity which brought their contents into existence was of a creative character or not. Seeing that in the course of our examination of the first function the elemental objects have already been taken stock of, we are now enabled to avoid the two grossest shortcomings of the current definition by simply referring to our earlier findings.

The second question to be answered by the definition of the second function concerns, I repeat, the principles governing the selection of those particular elemental objects which have to form an individual composite object of this function. Our examples illustrating the scope and limits of the individual objects of the second function as determined by actual practice, have shown that the limits within which cataloguers in fact do regard and treat books with different contents as components of the same object of this function, i. e., as editions of the same work, are very far from being self-evident; they must be determined, therefore, very carefully.

For large research libraries, which we are concerned with, the limits of the individual objects of this function have to be drawn more liberally than for public and small libraries. Accordingly, in view of the rather generally adopted practice of these libraries, the definition of the second function might be given the following general form: The second function consists in forming composite informational items out of the catalogue entries of such elemental objects the non-material constituent part of which is either a particular original text or other document-content, or else its intellectual descendants, in so far as they are likely to be considered interchangeable by a reasonable number of readers.

This definition shares a formal merit with the current one: it represents the second function as a homogeneous task which remains in all individual cases qualitatively the same. It achieves this by not distinguishing between the diverse kinds of the components of the objects of this function, i. e., by regarding them as a homogeneous mass. To this merit it adds, therein favourably differing from the current definition, the merit of not contradicting actual current practice. But, alas, the concept 'descendant of a work', and the phrases 'a reasonable number of readers' and 'likely to be considered interchangeable', which it contains, are much too vague; and so, as to clarity and precision, the above definition, too, lags very far behind the standard which cataloguing codes and theoretical argumentation must necessarily reach.

58. A PROPOSAL FOR AN APPROXIMATELY CORRECT DEFINITION OF THE SECOND FUNCTION. Therefore, if one wishes to complete the sketchy current definition of the second function so that it should attain a satisfactory degree of precision and definiteness, one cannot avoid founding it, instead of on a conception which represents the components of its objects as a homogeneous mass, rather on a less abstracting and generalizing way of looking at them. Such a more adequate definition would have to be based on a specification of at least all the principal types of components which, according to our present-day notions, the individual objects of the second function should comprise. Though from the point of view of logic only moderately satisfactory, this seems to be, nevertheless, the optimal expedient, seeing that it enables us to attain a higher degree of veracity, to approach the reality of present practice more closely, than any other formally more accomplished formula might do. The price to be paid for this advantage is that the validity of such a definition is liable to be less long-lived than that of a definition not based on a specification.

The completed version of our definition we propose then to formulate in the following manner: The second function of the author-title catalogue consists in providing pooled information about all those elemental objects recorded in it, the non-material integral part of which is either the original version of a particular text, collection of texts, or other document-content, or else a revised version, an enlargement, abridgement, adaptation of, or selections from, this original version – whether in the latter's language or in a translation.

By way of interpretation we must restate here that a physical unit, e. g., a book, may contain several elemental objects, if the parts of its contents – for instance the work that forms its nucleus, the commentaries attached to the latter, an introductory study, etc. – have formal marks of their own, and are recorded in the catalogue autonomously, under these formal marks. All these parts of the contents of a book form components of as many diverse objects of the second function. The same applies also when there are two *primary* elemental objects amalgamated in a single physical unit. A multi-volumed collection or a series itself is – in principle always, and sometimes practically, too – also a component of a particular object of the second function; while all the separately recorded members forming this collection or series are components of an equal number of objects of this function, different from the former as well as from each other.

The final form which we have given the definition of the second function does not fail to accentuate the material side of its objects either. Necessarily so, seeing that we have had to aim at a practically employable formula, and that in practice the catalogue-worker has always to deal with books, i. e., with items which have also a material side. Nor is the current formula able to refrain from accentuating this aspect. The fact that in defining the second function one cannot avoid considering the material aspect of its objects, has no bearing at all upon the validity of our former statement, according to which the generally adopted mode of cataloguing these objects eclipses in a certain sense their material side.

Occasionally it may be found desirable – e. g., in introductory paragraphs of codes – provisionally to substitute for the above elaborate formula a more concise and striking one. For this purpose I propose the following formula: The second function consists in presenting pooled information about all the components of each available individual literary unit. To the term 'literary unit' we manage to impart a clearly defined meaning: for us it denotes the aggregate of the elemental objects connected with each other by their contents in the ways specified in our more elaborate formula for the definition of the second function. So it is used here with an emphasis somewhat diverging from the usual one. As generally applied, the term seems to consider only the intellectual aspect of the objects of the second function, i. e., their common intellectual side; but it does not stress the fact that it is always elemental, material, objects which constitute them.

The word 'work', contained in the current formula, must in this context be discarded entirely. It would be very inconvenient if we should have to abstain from using this word in its loose everyday sense (or more precisely in the second of its relevant everyday senses) even in cases where the exact limits of the scope of the second function are of no consequence whatever. So, seeing that we must not go on using it in two different senses, we must, evidently, cease using it as a technical term, instead of 'composite object of the second function', or simply 'object of the second function'.

I think, our form of the definition may be considered a sufficiently accurate and articulate expression of the real intentions of the majority of present-day cataloguers concerning the scope and limits of the second function. Not quite assured about it, I only hope that I have not overlooked any of the really significant types of constituents which appear already in these days in the composition of the objects of this function.

From our definition we may infer that as to the degree of their complexity, there are very great differences between the individual objects of the second function. Among the components of some of these objects all the types enumerated in the definition are represented by several items; whereas at the other end of the scale one finds objects all the components of which belong to a single type, and – if we accept the extensive interpretation of the term 'composite object' given above[50] – even such that are constituted by a single item of a single type only.

[50] See § 19.

59. THE DIFFERENT TYPES OF COMPONENTS OF THE OBJECTS OF THE SECOND FUNCTION. Our next task consists in supplementing the foregoing discussion of the general scope of the second function by a survey of the types of the components of its objects. The relevant fact we have here to deal with is that, though the great majority of elemental objects can become a component of only a single object of the second function, there does exist also a problematical minority which belong to two or even more diverse, though related, such objects. To this minority we have to pay special attention.

The elemental objects with entirely identical contents but diverse physical marks do not raise any difficulty; quite obviously they have to be treated as components of the same particular object of the second function. This applies also to the original text and its simple translations. Difficulties begin to arise only when the contents of the possible aspirants to a particular componentship display differences other than a diversity of language.

One of the bibliographical constellations involving such difficulties is constituted by the re-edition of a particular text (or other document–content) enlarged by a new preface, commentary, explanatory study, notes, vocabulary, etc., written by persons other than the author of the original edition. More precisely, it is constituted only by a part, a minority of such re-editions: by those the outward forms of which strongly emphasize an accompanying commentary; their majority do not raise any problem either.

As an example for the problematical cases let us take the editions of *Beowulf*. Obviously, the different editions displaying several diverse forms of the title of the epic, must be treated as components of the same composite object of the second function, in principle irrespective of all additions and appendices they contain. This also applies if the extent of the latter surpasses the length of the text of the epic itself. At least, it applies, if the author was modest enough to compose a title-page with, say, the following wording: *Beowulf. Accompanied by an analytical study on the language of the epic, by X.Y.* But what if he gives the title-page of the very same book the form *An Analytical Study on the Language of Beowulf, by X. Y; accompanied by the complete text of the epic?* Surely, many cataloguers will now enter the book under X.Y.'s name. This implies the appearance of a new primary elemental object among the components of an object of the second function: of the study by X.Y. For the text of *Beowulf* most cataloguers taking this choice will make an added entry, but some might omit even that, for instance on the formally logical grounds that they consistently ignore all appendices, not making even added entries for them. In the first case the epic itself, though reduced to the status of a secondary elemental object, invariably continues to be recorded as a component of the very same object of the second function of which also the editions *entered* under the uniform title of the epic form components. In the second case the epic itself would come to be dropped altogether, quite incompatibly with both the first and the second functions.

A problem slightly similar to the one arising in the former case the cataloguer encounters when dealing with revised editions prepared by a person other than the author. The majority of these he enters without hesitation under the same formal marks as the respective original editions; mostly preparing also an added entry under the reviser's name, which may alternatively be considered a tertiary or a co-ordinate added entry. If, however, the thoroughness of revision transgresses a certain limit, it becomes inexpedient to enter the revised edition under the formal marks of the original edition. It would be a mistake, for instance, to enter the latest edition of a medical handbook, which has been published in the course of half a century in a dozen editions and revised several times, perhaps even by different persons, by whom the original author was becoming gradually eclipsed – or else a film script based upon a novel – as new editions of the respective original works. The advisable course is to enter such variants as entirely new works, forming components of objects of the second function, different from those to which the respective original works belong. But a cataloguer taking this course would make an unforgivable mistake, if he omitted to make an added entry under the formal marks of the original edition, in other words, if he omitted to record the new edition in its capacity as a component of that other composite object of the second function which is formed by the original edition and its slightly altered early revisions. Of course, also the earlier, less radically revised editions are components of two different objects of the second function, but in their case the revision is always recorded by means of added entries. The objects proper of these added entries constitute, together with the objects proper of the main entries of the later editions made under the name of the reviser, a new composite object of the second function, different from that other composite object which is recorded under the name of the original author, and of which the earlier components are recorded by main, and the newer ones by added entries.

The third species of the books under consideration consists of the separate editions of such relatively self-contained single parts of a work, or of a collection usually considered and treated as an integer work (e. g., the *Bible*), which have a title and sometimes also an author of their own, different from the title and author of the work of which they form a part. If such a part has been published separately only once or twice, these editions are usually treated as mere selections from the whole work, and entered under the latter's formal marks. They may or may not be accorded added entries under the formal marks of the part. But it would not occur to any cataloguer intent on fulfilling the second function of the catalogue, to make the main entry of such an edition under the formal marks of the part. If, however, the number of these separate editions increases, he will change his mind and begin to devote the main entry to the part, i. e., to treat it as a self-contained work, an independent literary unit. An example: Gottfried Keller's story *Kleider machen Leute*, which forms an integral part of his collection *Die Leute von Seldwyla*.

But if a cataloguer chooses to treat a part of a work in this way, as if it were a complete, integer work, he should at least never omit to prepare added entries (and only in extreme cases the explanatory references prescribed in § 124B of *AACR*) recording its editions also in their capacity as editions of a part of the complete work, that is to say, to provide access to them under the formal marks both of the whole work and of the part. This is quite indispensable to discharging the second function adequately.

The three catalogical situations just discussed are closely related. All the books displaying any of them have to be recorded in the capacity of two different elemental objects, either of which forms a component of such a composite object of the second function which differs from that object of which the other is a component. This trait they have in common with many other kinds of books, with many catalogical conditions. What distinguishes them from the latter is that the books displaying any of these three situations may be arranged so as to form a continuous scale. At the initial end of the scale it is quite indisputable which of the two elemental objects is to be considered and treated as a primary elemental object and allotted the main entry, and which as a secondary or tertiary one, or even one that can be left unrecorded. The first object, however, gets gradually displaced by the second, so that at the other end of the scale it is unequivocally relegated to the status of a secondary elemental object, and what is more, now it is this object which occasionally becomes entirely suppressed.

The cataloguer and code-maker must take special care to prevent that the latter alternative should come true. They must never omit to provide for a secondary entry being made under the formal marks of the commented work, the original work, the complete work. It is only in this way that they can avoid that important composite objects of the second function should be recorded only incompletely, and elemental objects absolutely worth recording suppressed entirely.

We may mention in passing also another problem raised by these three situations. The limit where the originally primary elemental object turns into a secondary one, and *vice versa*, is with them always liquid, and cannot be fixed by a rule of general validity. The only instruction of a general character which the cataloguer may be given in this respect is that he should obey the directions given by the wording of the title-page, and allot the main entry within each pair of objects to that object which the title-page emphasizes more strongly. This means, of course, that in these particular instances he should entirely resign himself to the domination of the outward, physical forms of elemental objects, i. e., to an often arbitrary lead. There seems, however, no better choice to be available. Especially, if one remembers that, first, the physical forms of elemental objects do, after all, determine the anticipations of a considerable number of the users of the catalogue, and second, that this is the clearest and strictest instruction that the cataloguer may be given, granting him the least possible measure of discretionary power.

The remaining types of the components of the objects of the second function – the abridged editions of, and the collections of selections from, an original – raise no problems. Our fundamental statement applies here too: such an edition cannot become a component of the corresponding composite object unless it is recorded by the catalogue as an elemental object. For instance, a selection of a poet's letters published as an appendix to a biography of him can become a component of the composite object formed by the editions of his correspondence only if the appendix in question *is* recorded autonomously. Conversely, if this appendix *is* recorded autonomously, if it *is* made an elemental object, then the cataloguer must not fail to make it also a component of the corresponding object of the second function.

60. THE FIRST AND THE SECOND FUNCTIONS ARE DISTINCTLY SEPARATE FUNCTIONS OF THE CATALOGUE. We come now to our final task connected with the second function: to examine the three questions as to the standing of a function among the tasks of the catalogue which – as we have indicated in § 11 – call for an answer with regard to each function. We have now reached a stage in which in relation to the first and the second function a part of these questions can be answered without difficulty on the basis of our foregoing findings.

Let us begin with the second of these questions: whether it is right to consider the task which we are calling the second function as a separate and distinct function of the author-title catalogue; and whether it would not be more justified to regard it as only a part of a more comprehensive function.

Let us see first whether the tasks we have called the first and the second function do not form between them merely a single function.

I can think of an argument in favour of an answer in the affirmative which might at first glance appear convincing enough. We had better, perhaps, dispose of it before we set out to examine the arguments in favour of the opposite, the correct, answer.

For cataloguers as well as for readers – so runs our tentative argument – a work is always only one of the two integral parts of a book or other document; neither cataloguers nor readers need ever deal with a work detached from all books; work and book are for them simply two different aspects of the same indivisible object. This is reflected, among others, in the age-old usage of allotting to work and book together only one single description, never two separate ones. It is true that this single description is often entered under two different headings and under two or three different sets of formal marks, at two or three different points in the catalogue; but this is by no means a sufficient reason for considering what we have called above the first and the second function as the latter's two separate and distinct functions. The device of entering the same book under two or more different headings is applied in general with the purpose of helping readers who are not informed adequately about the book they seek, or about the rules governing the arrangement of the catalogue. Multiple entries are made as a rule for the sake of fulfiling the functions of the catalogue with increased efficiency. Why should this not apply also to those instances in which a book is made retrievable under two different forms of its title or under two different names of its author? But if it does apply, then it makes no sense to speak in these instances of two different functions, to make a complicated matter out of a simple one without any reason.

This argument is obviously untenable. The analogy it is founded upon is false; and the resemblance between the two kinds of instances of employing multiple entries in cataloguing a single document is only superficial. It may, of course, happen that, for instance, an entry under the title of a translation may sometimes serve a reader looking for the original work but knowing only the title of the translation, or *vice versa*. The entry in question may perform this service, however, only incidentally, and therefore – as compared with its performance of the second function, i. e., with its being the means of conveying information, and of satisfying demands fundamentally different from those conveyed and satisfied by the first function – one can attach to this service of it only third-rate importance. This is, indeed, one of the rare points at which the cataloguers' motives for taking a certain course of action are completely clear and uniform: by the preparation of multiple entries of the sort we are talking about modern cataloguing practice aims quite certainly at meeting two or three different sorts of demands.

Having met this possible objection, we may now proceed to sum up the arguments, yielded already by our previous discussion, apt to demonstrate the validity of the thesis that the two functions in question are in fact separate tasks quite independent from each other.

We have seen that these two functions aim not only at essentially different objectives, but at objectives in some respects directly antithetical to each other. The first function consists in communicating information, based on their physical characteristics, about the single items contained in the holdings, regardless of any relationship existing between the intellectual constituents of these items; it presents the holdings in principle as an atomized mass of unconnected, detached units. On the contrary, the second function consists in disclosing a certain kind of essential relationship, rooted in their contents, between the items that are atomic only when recorded in behalf of the first function. No less antithetical are the respective natures of the objects of the two functions; although the cataloguer is dealing in both instances alike with single books, in the case of the first function it is the material side of these books which is his or her chief concern, in that of the second function their intellectual side, their content element. Moreover, the objects of the first function are simple in their structure, constituted by only a single item, whereas those of the second function are composite. These antithetical properties of theirs are reflected also in the sharply divergent character of the means employed in performing the two functions. Indeed, all these differences disclose the second function as a task more closely related to the fundamental function of the diverse (classified or alphabetical) subject catalogues than to the first function of the author-title catalogue. So there cannot be any doubt that one must recognize what we have called the first and the second function as two in fact entirely distinct and separate functions of the author-title catalogue. For a last argument we might also mention that their parallel performance generates continuous collisions between them, making even present-day cataloguing practice in many respects self-contradictory, often compelling cataloguers to adopt dubious solutions, and involving them in their most heated controversies. This would be entirely impossible if the two tasks constituted a single function, which would, of course, imply a logical unity.

In the foregoing we have answered the first half of the second question concerning the status of the second function. To complete our answer to this question we ought to examine whether the task which in this study we keep calling the second function does not form, together with the one we call the third function, a single function comprising both of these tasks. The examination of this question must, however, be postponed until we have finished scrutinizing the latter task. We may, however, mention in anticipation that the answer to this question will also be in the negative.

61. THE FIRST FUNCTION TAKES PRECEDENCE OF THE SECOND THEORETICALLY, BUT NOT PRACTICALLY. In coming now to the third question concerning the standing of the second function – to the question whether or not it is right to grant the first function priority over it – we must first of all point out that it is undoubtedly the first function which is usually considered by librarians as the higher-ranking one, and that this view of theirs is in its turn obviously based on the opinion that the first function is more indispensable than the second. In our days it would very probably be rather difficult to find, at any rate in large research libraries, a librarian who, if compelled to reduce his author and title catalogue to performing only a single one of its present functions, would not decide in favour of the first function. Proposals for jettisoning the other two functions have been voiced now and again, but I cannot recollect ever to have heard of anybody advocating a discarding of the first function. In harmony with this, ever since librarians have made

explicit statements on the functions of the author and title catalogue, they have always mentioned as first the function which in these days we, too, habitually rank as such. It is, however, not difficult to account for this astonishing, since unusual, concurrence of opinions. Librarians must have recognized that their work in general,and in larger libraries indeed even the maintenance of outward order, would become much more difficult if they did not possess a tool enabling them to trace each elemental object, first of all each primary one, under its physical marks.

It is more difficult to account for the fact that, in spite of the general agreement on the priority of the first function, no concurrence of opinions has yet been attained concerning the choice of that particular form of entry which should be devoted to this function.

At first glance it seems obvious that to the first of the functions, the first of the forms of entry should be assigned, that is to say, the main entry, which is undoubtedly the first, the most efficient and valuable among the three forms of entry, on the grounds that in a great many author and title catalogues it contains ampler and completer information, whereas in no one less, than the other two. In spite of this, the practice of employing the main entry, instead of the first, in behalf of the second and third functions, is spreading in our days with gathering impetus. This practice is undeniably equivalent to relegating the first function to an inferior standing in the catalogue, and to allotting to the other two functions practically higher ranks. There are not a few librarians who maintain that this is a practice which flatly contradicts the theoretical acknowledgment of the priority of the first function, and must be considered, therefore, as logically indefensible.

They are, however, wrong; the habitual practice hits the nail in this respect very accurately on the head. The employment of the main entry for assembling the information concerning a literary unit is, in fact, logically as well as practically superior to the opposite policy. As we know, the second function consists in assembling the information about a number of elemental objects with related contents. The most efficient way of carrying this out consists in heading the *main* entries of these objects by uniform marks. The opposite solution, thechoice of added entries and references for the purpose of collocating all information about the components of composite objects would scatter just those entries which are most efficient in achieving this end, throughout the whole catalogue. Added entries on unit cards form naturally an exception to this. But as there are still many catalogues, even apart from the printed ones, that do not use unit cards, but radically abridged added entries; and as, on the other hand, it would not be expedient simultaneously to establish two different practices in these two kinds of catalogues – this exception cannot devalidate, or even weaken, our above argument. Moreover, as already mentioned, the majority of catalogue-users do not look for a particular book, but for a particular literary unit; which means that for the majority it is more advantageous if the main entry is allotted to the second, and not to the first function.

These are decisive arguments: the employment of the main entry in assembling the components of the objects of the second function is certainly preferable to the reverse method. After all, the first function does not involve any collecting of diverse items of information; it is confined, without exception, to conveying single detached items. Consequently, if the catalogue-user finds at his first move a not quite adequate added entry or a reference only, this is considerably less irritating to him in the sphere of the first function, than in that of the second. In the former he will reach his goal already at a second move, and will be much less annoyed than he is liable to be if compelled to gather in person quite a quantity of dispersed information about a single literary unit from several distant points of the catalogue.

The performance of the second function by means of added entries and references is, besides, likely to deceive the catalogue-user unless precautionary measures requiring much extra labour and expense are taken. The user looking for a given literary unit, but acquainted only with the physical marks of one of its editions, which differ from the uniform marks under which the added entries of, or references to, all its editions are brought together, will find in a catalogue established in this manner only the main entry (or entries) of the edition(s) bearing that particular set of marks which is known to him, but mostly no indication whatever of the fact that apart from this (or these) the library possesses several other editions. So he may quite justifiably presume that there are no other editions available, and, more often than not, leave without having got what he needs.

The only way of avoiding this misunderstanding is to file, with the main entry of a particular edition made under its own set of physical marks different from the uniform marks of the literary unit, also a 'see also'-reference directing the user to that uniform set of marks. If there are several editions with diverse sets of such physical marks available, the said reference must be made, of course, in each case. This method is disadvantageous for the library, because to secure exactly the same results it requires a greater number of entries than the reverse one. But this is a drawback which is dwarfed by the plodding and annoyance of the catalogue-user who is sent from pillar to post: first from the main entry of a particular edition to the uniform set of marks of the literary unit, and then – finding here, only or partly, either added entries which contain as often as not only inadequate information, or else intrinsically inadequate references – from here again to the main entries made under the other sets of physical marks, and dispersed throughout the whole catalogue. Besides, this method makes also the selection of the particular edition which would fit (as to its date of publication, as to its language, as to the commentary or notes appended to it, etc.) his needs best, often very laborious for the catalogue-user; so that he may shun the trouble, and be prevented, thus, from making the most satisfactory choice possible.

It is not easy to suppress the suspicion that those who refuse to assign the main entry to the second function cannot be aware of these drawbacks of the reverse solution. This applies also to those among them – constituting an overwhelming majority – who advocate the reverse solution only in regard to titles, but not to authors' names. They do not seem to be aware of the indispensability of the corrective just pointed out (which holds, naturally, also in the case of author entries, with the exception of authors with very few entries): of the fact that it is not enough to file added entries or references under the uniform title which direct the user to the main entries made under the various actual titles of particular editions, but that this step must be supplemented by referring also the other way round. At any rate, the codes adopting the method of heading the main entry with the actual, the physical, title, have all failed to point out the necessity of applying this corrective.

With the relegation, whenever the second function becomes patent, of the first function to added entries or references, i. e., to a practically subordinate position, our early predecessors have made, then, a lucky stroke. Although according to present-day notions the function usually called the first, ranks in importance and indispensability, without doubt, as first among the functions of the author-title catalogue; yet – since this function can be performed with reasonable efficiency even if the first place in the catalogue, i. e. the main entry, is not assigned to it, whereas it is impossible to perform the second function efficiently enough unless this place *is* allotted to it – the practice of relegating the first function to added entries and references cannot by any means be considered illogical. One cannot speak in this connection either of theory having lost its foothold in practice, or of practice having taken a theoretically wrong course, as some people do.

The apparent contradiction they are objecting to is resolved thus in rather a simple manner: as to the two first functions the question of priority cannot be answered unequivocally. The right answer depends on the standpoint from which the question is examined: theoretically it is undoubtedly the first function which has precedence and greater importance; but this is quite compatible with its having to content itself, for very good reasons, with playing only the second fiddle in practice. Obviously illogical and self-contradictory is, however, the position of those who unconditionally accept the use of uniform authors' names in the heading of main entries, but refuse to treat in the same manner the uniform titles of anonymous works, seeing that the arguments in favour of the former solution fully applies also to the latter.

62. THE SECOND FUNCTION MUST DOUBTLESSLY BE RANKED AS A FUNCTION. The answers given in the foregoing two paragraphs to the second and third questions concerning the status of the second function, include also a full answer to the first of these questions, so that we may now dispose of this question very quickly. We have seen that it is only in certain respects that the second function falls behind the first one in importance. So, if one does not propose to reserve the designation 'function' exclusively for the first function – and it would be rather difficult to give adequate reasons for doing so – it is quite impossible not to assign it to the second function.

There is, however, in this context one more question which must not be left unexamined.

Even if it is admitted that the second function is a task of the utmost importance, the performance of which is absolutely indispensable, it is logically necessary to raise the question whether the author-title catalogue is really the most suitable instrument by means of which it may be performed, or not. Does not the fact that the elemental objects to be brought together on behalf of this function may display various physical marks, that their relationship is not based on their material, their formal side, but on their content, on their intellectual constituent – furnish an argument in itself valid against entrusting the author-title catalogue with this function? After all, the structure of this catalogue rests essentially on a formal basis, and has not much to do with the contents of the objects about which it is conveying information. The adaptation of this structure to the requirements of the second (as well as of the third) function involves necessarily an impairment of its formal character, and together with this also of its logic and consistency. All of us know only too well that this is really so, and that the second function (together with the third) is responsible for a multiplication of the difficulties and costs of author-title cataloguing. Why not transfer, then, the task which forms now the second function of the author-title catalogue to the classified or the alphabetical subject catalogue? After all, the basic task of the latter – which consists in bringing together documents that are formally unconnected, but related to one another by their contents – is in fact similar to the essence of the second function. Why does not relieve then the author-title catalogue of this task, which is in some respects alien to its essence, and render it in this way less hybrid in character?

From the point of view of logic no reason can be discovered for refraining from this step, but an examination of the practical side of the matter reveals several such reasons. First, the author-title catalogue has already in its infancy, as we have mentioned, approximated to the fulfilment of the second function, and so readers have grown accustomed to turn to it for information about literary units. Second, in consequence of the considerably greater technical difficulties involved, first and foremost of those raised by the inevitably differing mentalities of co-operating cataloguers as well as of those superseding each other, both kinds of subject cataloguing require in fact considerably greater efforts and costs than author-title cataloguing in order to attain the same degree of consistency and reliability. Consequently, as to precision and dependability, both classes of subject catalogues lag at present still far behind the author-title catalogue, and therefore a transfer of the second function of the latter to either of the former would necessarily entail a deterioration of the service concerned. Finally, the performance of the third function already includes, as an accessory product, also a partial fulfilment of the second function, which can be turned into a complete one by means of comparatively moderate efforts and sacrifices. Therefore, as long as the third function of the author-title catalogue is retained, it would not be reasonable to transfer its second function to another catalogue.

CHAPTER VII

THE THIRD FUNCTION: THE RECORDING OF PERSONS' AND BODIES' 'OEUVRES'

63. GENERAL REMARKS. The task we have proposed to call the third function of the author-title catalogue is in several respects related to its second function. The two functions resemble each other in that both are concerned primarily with the intellectual aspect, the non-material constituent of their objects. Accordingly, the device generally employed in performing both of them is in principle identical, it consists of assembling under their respective uniform marks all the components of each of their single composite objects. In consequence, the third function, too, is very often merged with the first. It becomes patent only when an author is known under several names (including pseudonyms and pen-names) or forms of name, and the document in hand displays an author's name or form of name different from the one adopted as his uniform name; or else in the case of anonymous documents the authors of (or the contributors to) the contents are known, and recorded by the catalogue.

There is, however, also an essential dissimilarity between the second and the third function: the components of an object of the third function are less closely related to each other than those of an object of the second one. Apart from their common origin, from an, at least partial, identity of their authors, the components of an object of the second function are also related to one another by these contents, or at least by an identical nucleus of these contents; whereas a great majority of the components of the objects of the third function lack this second bond. In consequence, the systematizing and bibliographical trait in the character of the third function is more apparent than in that of the second. The character of the third function is thus farther removed from the first function – which we have to regard as the most essential function of the contemporary author and title catalogue – than that of the second function.

Indeed, the remoteness of the character of the third function from that of the first is such a marked one that we cannot omit to raise the question whether the former is not intrinsically incompatible with the predominantly formal nature of the author-title catalogue? From the point of view of logic one might answer this question in the affirmative, but – just as in the case of the second function – on second thoughts one sees oneself compelled to alter this answer. The reasons for this are, here again, practical. Firstly, the bibliographical apparatus at the disposal of the present-day librarian frequently fails to provide information on an individual author's entire production in a really expedient, easily accessible manner; and secondly, all the other instruments produced by the libraries themselves are less suited to conveying this class of information than is the author and title catalogue. So, in spite of all adverse arguments, the generally adopted practice which allots this task to the author and title catalogue is even in our own days still preferable to any other solution.

In analyzing the third function we shall have to deal, on the whole, with the same tasks we have also been confronted with in the case of the first two functions. We must examine first the scope, i. e., the objects, of this function; and afterwards we shall have to answer the three questions concerning its status among the tasks which the catalogue should fulfil. This time, however, we shall already be able to utilize some results reached in the preceding chapters; a few of our former findings will be applicable to the third function even without any modification worth mentioning.

64. THE SCOPE OF THE THIRD FUNCTION IS NOT CONFINED TO A RECORDING OF ONLY AUTHORIAL ACHIEVEMENTS. According to the current formula the third function consists in conveying appropriately pooled information about the whole 'oeuvre', i. e., about all those editions of the works of any particular author which are contained in the library. Hence, the objects of this function (which, I repeat, is short for "the objects in the recording of which this function consists") are composed of that part of the materializations of individual authors' 'oeuvres' which is available in the library. This is a rather sweeping statement which we must, naturally, concretize. Its wording suggests that this we may achieve by analyzing the denotations which the words 'author' and 'oeuvre' bear in both the current formula and in this statement. This analysis will be split here into two parts, by being carried out separately first for personal authors and afterwards for corporate bodies. Our findings will show that such a separation of the two topics is not only justifiable, but logically cogent. Let us begin with an examination of the personal sector of this function.

The word 'author' in the current formula implies that the third function is concerned only with such productions of a person which may be classified as 'authorial'. The scope actually assigned to this function by practice is, however, a much wider one. It includes along with the authorial productions also all other achievements and performances recorded in the catalogue under personal names: in addition to the authorial productions constituting the sole or the predominant content of a book, not only the less self-contained authorial productions (e. g., forewords, epilogues, commentaries to other authors' works, etc.), but also all non-authorial works and acts a person executed, e. g., in the capacity of translator, of editor, or even of an issuer of decrees or ordinances, or in that of a sponsor. If in a particular catalogue a performance of any kind has obtained the status of an elemental object, and is recorded under the name of a person, it becomes automatically a component of the object of the third function concerned with that person. So this function is by no means confined to procuring information about authors' works only; its scope, and also the corresponding denotation of the word 'oeuvre', are much more comprehensive than that.

Both are, however, bound to vary almost from library to library according to the changing liberality with which the different libraries choose to record secondary and tertiary elemental objects. Therefore, it is impossible with general validity to determine the objective and *real* limits either of the catalogical denotation of the word 'oeuvre', or that of the scope of the personal sector of the third function, although we do not encounter any difficulty in constructing a generally valid formula for determining both these limits *formally*. With respect to each particular author-title catalogue they are quite unequivocally determined by the range of the elemental objects recorded in that catalogue under personal names. The only possible really valid and correct definition of that sector of the third function is to be based, therefore, on this formal delimitation of its scope, in the following manner: the personal sector of the third function consists in collecting all the separate single units of information about the respective productions (and other performances) of each person figuring in the catalogue, so that they should form new, composite units of information, the object of each of which is the whole of a person's available production.

This definition jettisons, as superfluous, not only the concept 'author' but also the concept 'oeuvre'. This is an obvious merit, seeing that the current definition of the third function is rendered inaccurate, i. e., not corresponding to the real intentions of cataloguers, by its use of the word 'author', and seeing further, that the limits of the denotation of the word 'oeuvre' cannot be determined objectively, in consequence of their varying almost from library to library. A comparison of our definition with the current one shows that the latter is inadequate for just the same reason as both the current definitions of the other two functions; by indicating only one, though doubtlessly the most important one, of the tasks the third function comprises, this definition, too, remains markedly incomplete. It offers only an approximative characterization, a vague intimation of the nature of this function, and fails to determine its actual scope and limits with every possible accuracy.

It cannot be denied, however, that even the definition we have just given is not quite free from obscure points either. These become apparent when one attempts to draw a line of division between two contiguous individual objects of the third function which include components belonging to one of the three species of objects of the second function that have already put difficulties in the way of our defining *that* function – the works published together with their commentary, the works revised by a person other than their author, and the separately published parts of works or collections. These difficulties originated there from the circumstance that a single unbroken graded scale of elemental objects had to be cut up, in spite of its uninterrupted continuity, into two (and sometimes even into more) distinct objects of this function; and that the exact point at which the cut had to be made, i. e., at which a continuous quantitative change had to be considered as having turned into a qualitative one, could not be fixed on an objective basis, nor, in consequence, in a generally valid manner. In the case of the third function we must resign ourselves again, just as we have been forced to do in the case of the second, to the impossibility of eliminating these blind spots from its definition. So we cannot succeed in attaining an entirely adequate definition of the third function, either; but nevertheless, as to definiteness and articulateness, our definition is again a great improvement on the current definition. The methods that may be used to bridge over its remaining gaps in practice, are the same we have already indicated when dealing with the second function.

Since the third function may be defined, and its scope determined with the greatest attainable accuracy without any reference to the concept of authorship, in this study we have, as a matter of fact, no cogent reason to bother about the precise denotation of the word 'author'. We may recall that in dealing with the first function we have found ourselves in the same situation. It proved possible to take stock of all the three main classes of its objects and to determine their range, without being confronted with the necessity of clarifying the concept of authorship. The range of not one among the three classes of elemental objects corresponded to any acceptable meaning of the word 'author'. Part of the primary and secondary elemental objects, and all the tertiary ones, have nothing to do with authorship at all; and conversely, important sections of what are unquestionably authorial achievements remain outside the range of elemental objects, and do not pertain to any of their three classes.

65. Excursus on the catalogical term 'personal author'. Although irrelevant to our subject-matter proper, the concept of authorship and the exact meaning of this term are, nevertheless, very significant for cataloguing in another context, namely in connection with the choice of the heading of the main entry, in other words, with the fixing of the boundary lines which separate the fields of application of the three diverse forms of entry from each other. So I think I had better disregard its irrelevance to my subject, and discuss here, by way of a terminological digression, the denotation of the word 'author' relevant to cataloguing. I feel the more justified to do so, as this digression will offer me an excellent opportunity for promoting the second objective of this study, referred to in the preface.

The denotation which the word 'author' bears in common speech is clearly confined to personal authorship; the word applies principally to persons who have produced a literary, scientific, or other kind of work. We must emphasize the last word: by cataloguers the expression 'author' ought to be used always for designating the originator of a 'work' and not the originator of a 'book'; books having besides their authors also other co-originators (editor, publisher, printer). After having done so already in connection with the first two functions, we must here emphasize once again that cataloguers should never fail to distinguish carefully between 'book' and 'work', in order to avoid the unpleasant consequences of a negligent use of the two words we have already illustrated earlier (see pp. 96–97).

If several persons have played authorial roles in the production of a work, all of them must, evidently, be regarded as authors (co-authors, joint authors), even if there are scores of them. True, if they number more than three, most cataloguers do not allot them entries of their own, and their individual productions are not recorded autonomously in the catalogue. This does not alter, however, either the fact of their authorship, or the logical necessity of an extension of the denotation of the word 'author' so as to include them.

Books containing one work written by numerous authors, as well as those containing several separate works by several different authors, are often entered, with complete neglect of the real authors, under their editor's or compiler's name. In this situation the latter simply suppress the author; which means that the concept 'author' on the one hand, and the concepts 'editor' and 'compiler' on the other, present themselves as directly antithetical ones. This clearly shows that it is a grave mistake to invest the word 'author' with an extended catalogical meaning to include also the editors and compilers of the kinds of books just mentioned; for this implies that to such books two different kinds of authors are attributed,though only implicitly and inarticulately; which is impermissible terminologically, and liable to result in confusion in practice. Cataloguers cannot dispense with the concepts 'editor' and 'compiler', and therefore they cannot permit themselves to construe these and the concept 'author' as partly overlapping concepts, or to use terms with such overlapping meanings. Moreover, it is quite possible to draw a tolerably clear line of division between authors and editors, or between authors and compilers; but there is no possibility at all to draw such a line between editors and compilers who *are* to be included in a catalogical denotation of the word 'author', and those who are *not*. Yet, to sacrifice a concept and a term with fairly clear contours for such as can be given only heavily blurred ones, is one of the biggest mistakes a cataloguing theorist or a compiler of a code can ever make. One should obviously infer from this that the proposal to jettison main entries under editor or compiler altogether – made in Paris by A.H. Chaplin and H. Braun and backed by a slight majority of the Conference – was a happy one, advocating a method far superior to the widespread present practice sanctioned by tradition.

From the fact that it is a mistake to extend the denotation of the word 'author' to include editors and compilers it follows that in the definition of authorship one must replace the word 'produce', which permits such an extension, by the word 'write'. Let us, then, try this trivial formula: an author is a person who has written a work.

One might suppose that in general usage the phrase 'author of a work' denotes the originator of the thoughts contained in the work as well as the originator of the form, and even of the wording in which these thoughts are expressed. This does not hold, however, at all; according to the generally accepted meaning of the word, an author needs be neither the one nor the other. The thoughts expressed in a work may be altogether the thoughts of persons other than the author, as in the case of a textbook, or of a popularizing exposition, say, of Einstein's theory of relativity; and they may even be stolen thoughts: we are accustomed to consider and to call the popularizer and the plagiarist the authors of popularizing and plagiarized works. On the other hand, the inward or the outward form of a new version of a work, differing form the original in its structure and arrangement or in its wording and style, respectively, or even in both, may also originate with persons other than the author – as in the case of a thorough revision, an abridgement, a translation, etc. – without affecting the authorial status of the original author.

Now, the last form we have given to the definition of authorship does justice to common as well as to professional usage in so far as the thoughts expressed in a work are concerned: it does include the popularizer as well as the plagiarist. But it does not do justice to general usage with respect to the inward and outward form in which a work makes its actual appearance: it does not exclude definitely enough the reviser who altered the original structure of a work only moderately or the translator who is the originator of the new outward form, of the entire text of the translation, etc. So the definition is in need of a further modification which should make it clear that revisers, translators, and other contributors with secondary roles, having a part in the production of only a particular version of a work, do not come under the head 'author'. We recommend, therefore, as final the formula that *an author is a person who has written the original version of a work* (which calls, of course, for a complementary definition of the co-author). This definition makes it unmistakably clear that later editions, translations, revisions, and other variants of a work are to be regarded as having the same author as the original work, even if their new form, structure or text have originated wholly or partly with somebody else (who may in some cases also be treated as co-author).

This definition is in harmony with everyday language. It establishes a single criterion: the person who is to be considered the author of a work must be the originator of the text, of the form, of the original version of this work. In accord with the denotation which the word 'author' in common speech in fact bears, this definition does not make it a condition that the thoughts, the ideas, expressed in the work should be the thoughts or ideas of this person; nor does it exclude this possibility. It is thus in harmony with the generic nature of the everyday word 'author', which denotes, as a matter of fact, a large genus comprising many greatly differing species of phenomena.

The proposed definition harmonizes on the whole, with the generally adopted professional terminology and with cataloguing practice; but not absolutely. It does not seem possible to attain a definition of authorship to suit perfectly the needs of cataloguing. The above definition falls short of achieving this, first in consequence of its not settling definitely the pigeonholing of a few rather common kinds of works, and secondly because there exist some other kinds which a great majority of cataloguers are accustomed, and for more or less sound reasons they wish to continue, to treat in a way that is not in accord with that definition.

All these cases must be dealt with separately and individually; theoretically by means of an appropriate interpretation of the definition; practically by supplementing the general rule in codes concerning author entries – which should correspond to the above definition! – with appropriate specific rules. So it must be stated, first of all, that dictionaries, bibliographies and other compilations of data, e. g., collections of inscriptions or of tables of figures – although they do not contain a text 'written' in the ordinary sense, and are only 'compiled' by their originators – should be regarded and treated, nevertheless, as authors' works. We should also mention, as special cases in which our definition is contradictory to sound and generally accepted cataloguing practice, that many cataloguers are accustomed not to enter old dissertations under the respondent's name, even when it is probable that he, and not the *praeses,* has actually written the dissertation or composed the text of the theses; or to enter a narrative told and composed respectively by two different persons, not under the actual composer of the written text; or again, to enter a dialogue or an interview not under the name of the reporter, a libretto not under its author, but under the composer. The preservation of these practices is not essentially important, but it is not objectionable either, provided that they are explicitly stated to be exceptions to the general rule. Finally, there are the three unwieldy species of books we had already to deal with at some length in another context (the books containing a work revised by a person different from its author, those containing a work and a notable commentary on it by two different persons, and the parts of works or collections frequently published also separately under their own special formal marks, cf. pp. 102–103), which codes must separately provide for in consequence of most of the books of these three kinds having two authors with different roles, neither of whom must be left unrecorded.

By way of conclusion I had better, perhaps, repeat that this paragraph has been a mere digression, and emphasize once more that all the above-mentioned particular groups of cases are of interest only from the point of view of the rules for author headings which have, of course, to be based on a definite concept of the 'author'. To our subject it is entirely irrelevant whether the personal productions in fact recorded are recorded because they satisfy the criteria of authorship laid down by any definition and whether they are recorded on main entries or not.

66. THE DEFINITION OF AUTHORSHIP IN THE A.L.A. RULES. Let us supplement our excursus on the definition (for cataloguing purposes) of authorship, by an analysis of the definitions given by the *A.L.A. Rules.* It is worth while to undertake this analysis, since the *A.L.A. Rules'* handling of the matter is extremely well suited as an illustration of our second premise and of the carelessness with which questions of this kind have been treated also in modern times and even by the most prominent cataloguers, and since it will serve also as a useful introduction and transition to our next topic: to the discussion of corporate authorship. Perhaps I had better anticipate the objection that today it is no more warranted to bother about analysing the *A.L.A. Rules,* as since their replacement by the *AACR* they have already begun slowly to fade away into the past. I meet this objection by pointing out that the illustrative value of the *A.L.A. Rules* which I have referred to has by no means been diminished by their having been superseded by a new code, and that besides, only some of the mistakes contained in the passages we are going to analyze were discarded together with the old Code, while the most substantial ones among them have managed to hold their own also in the *AACR.*

The first two sentences of the very first paragraph of the *A.L.A. Rules* run as follows: "Enter a work under the name of its author whether personal or corporate. The author is considered to be the person or body chiefly responsible for the intellectual content of the book..."[51]

Let us for the moment disregard the fact that the introduction of the concept of corporate authorship, a concept unknown in common speech, in the very first sentence of the Code was an impardonable mistake (we shall come back to this topic presently), and begin by stating that, even apart from this additional obscure point, that sentence is seriously in need of an interpretation, first of all as regards the sense to be attached to the word 'author'. Codes as well as theoretical papers on cataloguing keep using this word in widely diverging senses, and so one cannot employ it in a rule without carefully specifying the particular meaning which one allots to it. The second sentence of the above citation does contain something in the way of an interpretation, but alas, not at all *the* interpretation, *the* definition, one is entitled to expect. As the first sentence refers to the author of a 'work', it is inexcusable that in the second sentence there should appear unexpectedly, as though by a conjurer's trick, the author of a 'book'. The definition it contains is, in consequence, no appropriate supplement to, nor an interpretation of, the first sentence. This slip in logic has, of course, lamentable consequences. The second

[51] *A. L. A. Cataloging Rules for Author and Title Entries.* 2nd ed. Ed. by Clara Beetle. Chicago, A. L. A. 1949, p. 3.

sentence undeniably implies that, for instance, so far as the 'book' in our former example (see p. 96) is concerned, Livy is not to be considered its author, and that, consequently, the first sentence simply does not apply to him!

This slip originated in the fact that while in the first sentence the compilers of the Code have been concerned with a fundamental problem: with stating that the first place in the ranking order of formal marks is the due of the author's name, which, of course, involved the obligation to define the concept 'author'; in the second they torpedoed in advance every chance of an adequate solution of this task by introducing, very inopportunely and prematurely, also another matter, and one of only secondary significance at that: the question of an extension of the field of application of author main entries beyond the limits fixed by the common meaning of the word 'author'. This intermingling of two entirely different matters has led to a deplorable distortion of the resulting concept of authorship. Moreover, it has also made, in conjunction with the accompanying inadequate mode of expression, both of the analyzed sentences misleading in certain respects.

The second sentence is apt to generate the false impression that it contains a definition of authorship which has in cataloguing, or at least for the whole Code, general validity. It certainly, however, does nothing of the sort; it contains merely an attempt, an abortive one at that, summarily to fix the boundaries of the application of author main entries. It is, obviously, only for the sake of this attempt that the first definition of authorship given in the Glossary appended to the Code, according to which an author is the writer of a work, had to be dropped this time. That simple definition could not be retained here, because what the makers of the Code really aimed at was a fixing of the boundaries of the application of author main entries in a manner entirely incompatible with every meaning which the word 'author' may bear in everyday language. They wanted to include within these boundaries the corporate 'authors', as well as some classes of personal editors and compilers, who are, obviously, *not* considered authors by anybody; and to exclude the authors of not independently published works, who *are* considered authors by everybody, even by all cataloguers. These were the reasons for the astonishing popping up in the second sentence of the word 'book'. The way which could have led to a fulfilment of the expectation, raised by the first sentence, of a generally valid definition of authorship, became in this way definitively blocked.

As for the first sentence, this is in turn deceptive, because the rule it contains has never been, and has never even been intended to be, put into effect except partially: only in regard to works that have not been published in collections and their like.

Furthermore, the second sentence fails not only to define in a generally valid manner the catalogical concept of authorship; it fails even to fulfil the second purpose (the determining of the boundaries of the application of author main entries) emerging in it so unexpectedly. The formula 'chief responsibility for the intellectual content of a book', chosen with a view to this purpose, is indeed a very unhappy one. By its vagueness it is bound irreparably to wreck any rule or definition in which it is incorporated. It is not only that it makes no sense to say that Aristotle is to be considered the author of *Organon* because he is chiefly responsible for its intellectual content. The really conclusive argument against this formula is that it is in every respect indistinct and therefore quite unfit to be used in a rule. The word 'responsibility' denotes a broad scale of not only quantitatively, but also qualitatively different things. The 'responsibilities' concurring in the creation of the intellectual content of a book may widely differ in quality, and be quite incommensurable, which implies that very often the 'chief' one among them can be determined only absolutely arbitrarily. A particular aspect of this difficulty we have already mentioned earlier: the line of division between those editors and compilers who are to be considered 'chiefly responsible', and those who are not, cannot be drawn satisfactorily. The same applies to the multifarious combinations of actually concurring personal and corporate authorships, editorships, compilerships. The cases in which the formula completely fails the cataloguer become, however, most frequent in the field of application of corporate main entries, i.e., in the very field for the sake of which it was mainly created. The absolute unsuitability of the concept 'chief responsibility for the intellectual content' not only of a book, but even of a work, as a catalogical criterion, manifests itself best in the fact that it may also become apparent in otherwise entirely unproblematical cases of personal authorship. According to this criterion it is obviously Einstein who is to be regarded as the author of a popular work containing a mere summary of his theory, and not the Mr. Smith who has actually written this work, but is really very far from being chiefly responsible for its intellectual content.

The practical consequences of the failure of the formula analyzed to determine the boundaries of 'authorship' vary very widely. In the region of personal authorship the formula is not liable to make mischief, because ingrained cataloguing habits dependent on the everyday meaning of the word 'author' prevent its implications contradictory to these habits from being noticed at all. But when it comes to treating editors, compilers and corporate bodies as authors, the practical consequences become rather deplorable.

The unlucky would-be definition in the second sentence we have analyzed, is also a source of secondary slips. Thus, for instance, the makers of the *A.L.A. Rules* did not hesitate to voice explicitly the implication of this definition that the personal as well as the corporate compilers and editors of books containing the works of many contributors have to be regarded as authors, which is from their point of view nothing but logical. At the very first opportunity of applying it, this new, comprehensive meaning they attached to the word 'author' was, however, already forgotten: by calling in § 5A1 an editor of this class 'editor', a compiler 'compiler', and not by any means 'author', they reverted again to everyday usage.

In his 1960 Draft Lubetzky already got rid of this terminological curiosity, he refrained from using the crucial word 'author' in two different senses; but the *AACR* restored the old practice again in its entirety.[52] In this astonishing fact we can discern the manifestation of two different truths: first, that a mistake once made is apt to generate secondary mistakes; and secondly, that with the majority even of contemporary librarians a pious regard for formulae with a venerable past

[52] *Anglo-American Cataloging Rules.* North American text. Chicago, A. L. A. See Note 2 on p. 9 and paragraphs 4A and 5A. – One might mention that the makers of the new Code did not even content themselves with using the word 'author' in these two sharply diverging senses; they also introduced a third denotation, having no roots whatever in tradition, which includes also 'translators', etc., that is, any person mentioned in the catalogue, irrespective of the nature of his or her performance or activity (see Note 2 to § 40 of the Code).

is still carrying greater weight than their desire for an adequate professional terminology and their respect for the axioms by which such a terminology would have to be governed.

67. 'CORPORATE AUTHORSHIP' IS THE MOST OBSCURE TERM IN CATALOGUING TERMINOLOGY. Our next task is to examine the concepts 'author' and 'oeuvre' with respect to corporate authorship.

The definition of 'corporate body' and the delimitation of the phenomena to be regarded and treated as corporate bodies, give the cataloguer very little trouble. He may regard as a corporate body every union or organization, whether permanent or ephemeral, which has had a part (except in the capacity of a mere publisher's or printer's) in producing an elemental object of cataloguing (including its intellectual constituent, its content), and which is in possession of what may be considered a name. His difficulties commence only when he addresses himself to the task of fixing the meaning of the term 'corporate author'. In spite of this – or, perhaps, just because of this – not a few codes go out of their way to define the concept 'corporate body', but do not even think of doing the same for the concept 'corporate author'.

As we have seen above, the denotation of the cataloguing term 'personal author' can be determined with rather satisfactory precision on the ground of the meaning borne by the word 'author' in everyday speech. Contrarily, in establishing the meaning to be attached to the term 'corporate author' one cannot take everyday language, to which this expression is entirely unknown, as a starting-point. This is not quite without advantages, cataloguers being given, thus, a free hand to invest the term with a meaning exactly in accord with the needs of practice which have actuated them to introduce it. They did not, however, make any use of this advantage up till now, they never contrived to devise other than quite vague definitions of the concept, and even vaguer delimitations of the term 'corporate author'.

What may be the explanation for their permanent failure?

The notion underlying the widespread adoption of the concept and term 'corporate author' by cataloguers is that it is the corporate bodies themselves which produce the contents, the intellectual constituents, of many of their publications, and that this particular side of their activity developed in producing such publications is closely analogous to the activity of personal authors. It is obviously on this basis that the compilers of the *A.L.A. Rules* have ventured to introduce the term 'corporate author' in the very first sentence of the Code, as though 'corporate authors' constituted simply one of the species of an easily defined genus 'author', a species closely related to the species 'personal author' of the same genus; and that they have made, as soon as in their second sentence already, an attempt at formulating a definition including both these species of authorship. We have already demonstrated that so far as personal authorship is concerned their definition is a failure; and here we come to stating that it is a failure in regard to corporate authorship, too. The criterion 'chief responsibility for the intellectual content of the book' is with respect to corporate authorship no less inadequate than to personal authorship. It not only gives no clue to the handling of a large part of those cases in which several bodies, or bodies and personal editors or compilers, have shared parts in the production of a book or work; but it also frequently fails the cataloguer confronted with the task of choosing between a title entry and a corporate one.[53]

It is, however, not only the *A.L.A. Rules* which did not succeed in defining the concept of corporate authorship and in determining adequately the denotation of the term. Not a single code throughout the world has managed to tackle this task satisfactorily. To prove this, we have to point out but two facts. The first is that almost every code which adopts this concept defines it, if at all, differently from most of the others, while those refraining from defining it explicitly, invest it implicitly with denotations different from those adopted by the others, which alone suffices to demonstrate the arbitrariness of all their respective choices. The second is that those of their rules which are supposed to determine or to imply these respective denotations are, apart from their divergencies, without exception also more or less vague and obscure in themselves, and so liable to generate fluctuating, inconsistent practices. All of this applies, curiously enough, also to the *AACR* in full measure. This code makes an effort to avoid, as far as possible, the use of the term 'corporate authorship', but without succeeding in palliating thereby the inadequacy of its rules for entry under a 'corporate body' that is 'the author', which on the whole are no better than those of the *A.L.A. Rules*. They display not only several rather conspicuous lacunae, but there are also many among them which are unsatisfactory as a result of a use of concepts with 'blurred edges'.

Both the inability of code-makers to agree on the meaning of the term 'corporate author', and their failure to state in an at least approximately exact manner their own respective versions of this meaning, are the unavoidable consequences of a lack of any objective basis upon which a generally acceptable definition, delimiting with satisfactory sharpness the range of phenomena to be subsumed under it, could be founded. This is a situation frequently occurring in cataloguing (and wherever the application of any kinds of rules is involved, for that matter), the reasons for which we had better examine more closely.

68. THE CAUSES OF THE PERSISTENT FAILURE TO DETERMINE THE DENOTATION OF THE TERM 'CORPORATE AUTHOR'. With the exception of individual names all words are generic, but there are great differences between their respective grades of genericness and abstraction. Among technological terms there are many which denote only a single kind of strictly congruous things. Such a term is generic, too, in the sense that it denotes a multiplicity of individual objects; but all of these objects are specimens belonging to a single sharply and clearly definable species, with regard to which the term is no more generic. The meaning of such a term is sharply outlined and for all practical purposes exact; the term itself is in the highest possible degree specific. The better part of words and terms, however, even of those belonging to specialists' terminologies, have a very different character: their denotations include several species, a whole genus of phenomena. This is the case with most cataloguing terms, too. For instance, the denotation of the term 'per-

[53] Cf. DOMANOVSZKY, Á.: Code-making – a Criticism and a Proposal. *Vjesnik Bibliotekara Hrvatske*, Zagreb. God. 14. Broj 1/2. Spomenica Evi Veroni. 1968, pp. 58–68.

sonal authorship' includes, as this was already touched upon above, not only in everyday language but also in cataloguing terminology, a rather broad scale of widely differing combinations of performances.

The world of those physical phenomena which the cataloguer has to deal with, the world of the primary elemental objects of cataloguing, resembles nature in exhibiting an almost endless variety of forms and combinations of forms, which one may conceive as arranged, in the same manner as systematic zoology or botany arrange the phenomena of living nature, into a graded system displaying a high degree of continuity; every member of the system being closely related to its next neighbours, and the less closely to the system's other members, the more its distance from them within the system increases.

Now, one is often obliged to allude to a particular portion of such a, one might say, three-dimensional continuum, to a portion that comprises still several phenomena which, though situated within a distinct segment of the continuum, not only differ from each other, but differ also in different degrees. The comprehensive name one will have to give to the whole set of phenomena the segment includes, will usually be a generic word belonging to the second of the above mentioned classes: a term the limits of the denotation of which it is not easy to fix, the segment having no clearly discernible and definable boundaries. The difference between two individual phenomena situated at opposite points of the periphery of the segment is, in consequence of the close relationship of all immediate neighbours throughout the whole continuum, usually considerably bigger than the difference between either of them and its respective neighbour which has its place already outside of the boundary of the segment. The drawing of the boundaries of such a segment is, thus, mostly not only an arbitrary undertaking, but also one which cannot be solved with precision and clarity.

In consequence, it is not easy to prevent the meaning of the generic word or term chosen to designate the sum of all the phenomena within such a segment, from becoming vague. Even in the physical world it is often difficult, and sometimes entirely impossible, to establish clear-cut border-lines splitting up a continuum into conceptually distinct parts (e.g., to determine the exact boundaries between the constituent colours of the spectrum). With the concepts we have to deal with, with abstractions and generalizations like 'corporate authorship', which have no counterparts in the world of reality, these difficulties by no means diminish. Wherever one may choose to draw the boundary lines of such a concept and term, there are liable to remain more or less notable contingents of cases in the world of corresponding reality, as to which it is dubious and open to dispute whether they should be subsumed under that concept and term or not. The meaning of some of the terms belonging to this class may appear at a first glance clear enough, and only their application in practice discloses that their denotation is in fact vague, that people keep differing about the classification of a lot of border-line cases, and that it is impossible to find an objective basis on which to settle the issue in a generally satisfactory manner. The edges of a great part of the generic concepts and terms of this sort are bound persistently to remain "blurred".

Let us illustrate this by a simplified, one-dimensional example. White and black are diametrically opposed qualities. Yet if you carefully arrange, according to their respective shades, a few scores of differently coloured men in a row leading gradually from the whitest man on earth to the blackest, and are faced with the task of dividing the row into a white and a black part, you will not be able to determine, in a manner acceptable to everybody or even to a majority, the exact point where white turns into black.

Against this example the objection might be raised that finding a valid way of cutting the said row into two distinct parts is so difficult only because individual tastes, interests, and prejudices are involved to a great extent; and that optics is capable of objectively determining a median, the exact middle between white and black. This admitted, we must conclude that for the drawing of the dividing line between those corporate bodies that are to be considered and treated as 'authors' and those that are not, theory and codification have not succeeded yet in discovering such an objective method; that is to say, that in this particular respect they have not yet outgrown the stage of founding their distinctions and judgements on individual tastes and prejudices.

69. It does not seem possible to overcome the difficulties of giving the term 'corporate author' a clear-cut denotation. The analogy between personal and corporate authorship is a mere fiction, and the application of the designation 'author' to a corporate body is nothing more than a metaphor. A personal author actually writes his work; he or she composes its text in person. As we have seen above, it is just this circumstance which is the only workable criterion of personal authorship, fit do determine its limits, the dividing line between authors and non-authors. Since corporate bodies are unable to write and compose and all their publications are written by persons, it is obvious that the criterion of personal authorship is unfit to serve as a basis for the definition of corporate authorship. This is the main reason why *all* attempts at amalgamating the two definitions – and not only the one made by the *A.L.A. Rules!* – are doomed to fail. True, many of the publications concerned are in fact written in the name of a body, vicariously, and the application of the term 'corporate authorship' is restricted by many codes to such publications only. It is however, quite impossible to draw a sharp line of division between those publications of corporate bodies that *have* been written vicariously, on behalf of the body, and those that *have not*; even irrespective of the rather common publications of which half of the contents has been written vicariously, the other half not. On the other hand, in many cases when the publication was undoubtedly written on behalf of it, a body is none the less not considered bv cataloguers and codes its corporate author. This hig-degree liquidity of its boundaries is the crucial point which renders the concept of corporate authorship definitively unserviceable.

A publishing firm which publishes the work of a well-known novelist is not considered by anybody a corporate author, since it did not contribute anything to the intellectual substance of the novel. At the other end of the scale (e.g., in the case of an annual report of a body) one is confronted with a diametrically opposed situation: the persons who participated in the composition of the report remain in the background and very often also anonymous; the report is the product of co-operation between several persons acting not on their own account, but on behalf of different organs of the body. In such a case it seems not unwarranted to speak about an authorship of the body; and it is also in this sort of cases of corporate authorship *par excellence* that one has to look for the reasons for the world-wide adoption of that concept and term. But that novel and that report are not typical for all corporate publications. They are the opposite poles – not only, as in our example of white and black men, of a single scale of continuously and steadily graded instances, but – of a wide field of extremely multifarious forms generated in abundance by the different agents having a part in the production of books. These forms display in practically every direction, as it were in three dimensions, the same gradual transition, the same slightness of the respective differences between two neighbours, which we have met with in our white-black scale. Setting out from the centre of the field to be covered by the term 'corporate author', i.e., from the particular phenomenon subsumed most readily under this term, many rows of gradually changing phenomena, resembling our white-black row, radiate in all directions. There is nowhere a clear dividing line discernible which might cut the whole range of corporate publications not only plausibly but also clearly into two opposite groups – into those which may be considered as having the body for their author, and those which may not. In other words, there is no possibility of finding any unequivocal criteria of corporate authorship, and of defining this concept, of drawing the boundary of the denotation of the term, in a not only substantially, from the viewpoint of the informational tasks of the catalogue appropriate, but also in a formally adequate, sufficiently clear and definite manner.

And then, there is still another difficulty. The middle between white and black may not only be determined with exactness optically, but the result can also be applied practically, simply by equipping everybody concerned with an appropriate optical instrument. There is no possibility of a corresponding device being applied by cataloguers in diagnosing corporate authorship, the situation being here, as we have pointed out, much more complex.

To sum up: I cannot see any possibility of formulating rules for the application of corporate main entries in a manner that could prevent an inordinately frequent occurrence of misunderstandings and misinterpretations of the intended range of their validity.

70. IT WOULD BE THE BEST DEFINITIVELY TO DROP THE CONCEPT AND TERM 'CORPORATE AUTHOR'. If, however, there is no such possibility, this ought to induce codifiers to drop all rules based on the concept of corporate authorship, and theorists to drop the concept altogether. This follows from what, I think, we may consider an axiom: both should refrain from adopting concepts, terms and rules covering only a portion of such a continuous and evenly gradated scale or field of phenomena as we have been speaking of in the last paragraph, if there is no possibility adequately to fix the limits of that portion. They may adopt such a term if the scale or field in question exhibits an appropriately located dividing line which not only facilitates the fixing of the boundary of the denotation of that term with sufficient precision and clarity, but also enables them to impart to the cataloguer and the catalogue-user a satisfactorily lucid idea about the location of this boundary. We must not forget that using a term with a vague meaning of the above type is an even graver mistake than that one may commit by using a word which has various related but still distinctly different meanings, without indicating that particular meaning one actually intends to attach to it. In the latter case the context may help the reader to find out the right one among the different meanings; whereas in the case of a vague term between the various possible denotations of which there are

only dim differences of degree, the context necessarily fails to offer such auxiliary guidance. The use of inadequate terms belonging to the latter class must, therefore, not only inevitably result in divergences between the cataloguing practices of different libraries, but very likely also in inconsistencies within the same catalogue.

The very minimum of terminological tidiness one must insist on is that theorists and rule-makers should refrain from employing vague concepts and terms of the above sort at least when they are dealing with questions of the heading and the entry word of main entries. Statements and rules concerned with these should be in the highest possible degree precise and unequivocal, seeing that the main entry is not only necessarily unique, but also the foremost, the most valuable one, since the richest in information, among all the entries referring to a particular object. If a concept of the 'blurred-edged' type is in spite of its shortcomings considered a cataloguing device so useful as to be indispensable, it is possible to make use of it by means of added entries. With an added entry one can always resort to the expedient of providing for its being prepared also 'in case of doubt', since a supernumerary inappropriate added entry cannot do much harm. Not so an inappropriate main entry, since by definition the main entry cannot be supernumerary. Therefore statements or rules concerning main entries should not contain the 'in case of doubt' clause; they should be drawn up in a clear and definite manner excluding even the possibility of a doubt worth mentioning. The cogency of this precept increases, naturally, in proportion to the importance and to the degree of generality of the matter dealt with by the statement or rule.

But is it not entirely impossible even to consider discarding the corporate main entry, and, indeed, the very concept of corporate authorship? Would such a step not entail an impoverishment of our thinking about cataloguing, and end in a diminution of the efficiency of the author-title catalogue?

It is first of all the American librarians who will be inclined to think so, seeing that they are accustomed to regard the rules for corporate authorship as their country's principal contribution to the armoury of author-title cataloguing. It must be admitted that there is some truth in this view of theirs, but one that needs qualifying. Their really valuable contribution consists in their having been the first to exploit *all* the possibilities of a utilization, as a handle of great worth, of the name of the corporate bodies which have had a part in the production of books or works; in their having recognized the aptitude of these names materially to improve the retrievability of a large and important class of objects of cataloguing; in other words, in their discovering the great value these names are capable of attaining in the role of formal marks. The merit of their having made this discovery may be set in the true light only if one contrasts it with the astonishing failure of another great cataloguing tradition, the German one, to appreciate this inestimable kind of formal marks. But at this point the merit of American cataloguers ends; most of their steps taken for exploiting the possibilities offered by corporate names – the introduction of the concept of corporate authorship, apparently for the sake of a subsequent rationalization and justification of their actual application of corporate main entries; the distinguishing, for the same reason, between two classes of bodies: those which are to be treated as authors of the books or works in the production of which they participated, and those which are not; and finally, the developing of an increasingly intricate and very artificial system of the application of corporate main entries (which included the unnecessary and unfortunate move of dividing the 'author'-bodies in an extremely nebulous way into three different classes and allotting to each class its own special form of heading) – these steps have been very unlucky, and their outcomes have thrown a great burden upon author-title cataloguing throughout the world. This cannot be denied by anybody who dispassionately considers the share of faulty corporate author entries in the total of the mistakes made by cataloguers, or else the share of the complaints made by readers about corporate entries in the total of their complaints against the author-title catalogue.[54]

There is another curious circumstance which we cannot abstain from mentioning before concluding this criticism of the conception of corporate authorship still in vogue today. The fact that most codes do not contain any definition of corporate authorship implies that the meaning they attribute to this concept, and the denotation they assign to the term, must be deduced from their concrete – and logically as a rule rather objectionable – rules specifying the publications which are to be given a main entry under a corporate heading. This means that they neglect to give their pertinent rules a general theoretical, conceptual foundation, and leave it to the cataloguer to bring some order into his own ideas about this matter, if he can. This appears to me a logically impermissible, and practically inadequate way of proceeding.

71. The real range of the objects in the corporate sector of the third function. In the foregoing paragraphs we have dealt, with respect to the corporate sector of the third function, with the second of the two tasks of this study. We have examined the concept and term of corporate authorship as applied by present-day cataloguing theory and practice. Now we have to turn to our first objective, our subject-matter proper, which here consists in a positive determination of the essence of this sector.

[54] But it is not only to cataloguing practice that the American conception of corporate authorship has given an awkward twist; as it was to be expected, theory too has come to bear some queer fruits under its impact. An outstanding one among these may be found in Julia Pettee's paper on The Development of Authorship Entry and the Formulation of Authorship Rules as Found in the Anglo-American Code (*Library Quarterly*, vol. 6.1936, pp. 270–290). The main train of thought of this paper is concluded by the sentence: "If the entry under title for these two groups", – i. e., the anonymous classics and all other works of personal authorship whose author is unknown – "may be considered as substitute for author entry, as I think it may be, all main entries may be considered authorship entries" (p. 286). To consider a title entry as authorship entry is equivalent to considering a thing black which is undeniably white. What may have been Miss Pettee's motive for doing this? The only one I can conceive of is the general preference for simple explanations, theories, representations, even where complex things and phenomena are concerned; a characteristic trait of human thinking. Miss Pettee's aim seems to have been to represent the undeniably over-elaborate system of American cataloguing rules as though it were based on a single simple underlying principle. In this striving after a beautifying representation she was only faithful to the traditions of American cataloguing theory. This had developed out of certain kinds of stray catchword and subject entries used at random by descriptive cataloguing (in the wider sense) in its primitive stages, three entirely new separate categories of main entries (the corporate entries made under the name of (1) a body, (2) a jurisdiction, and (3) the place where a body has its seat), and then, by subsuming all of these under the category of 'authorship entry', it has made an – abortive – attempt at reducing the number of the in fact applied numerous different kinds of headings to a mere two. No wonder that Miss Pettee could not resist the temptation of producing the missing link, and of substituting comsummate unity for the would-be duality already apparently attained.

Our definition of the personal sector of the third function is valid also for its corporate one, if for the word 'person' the term 'corporate body' is substituted. This sector consists in collecting all the separate single units of information about the respective productions (and other performances) of each corporate body figuring in the catalogue, so that they should form new, composite units of information, the object of which is the whole of a body's available production. Whether a particular performance of a body is of an authorial nature or not, is from the point of view of the third function just as irrelevant as the authorial nature of a personal performance; the editorial, sponsorial, translatorial, and any other possible performances of a body, if they are recorded autonomously, must also be ranged among the components of the object composed of the assembled results of the activity of that body.

From the irrelevance of the differences in kind between the performances of bodies which may form parts of a 'corporate oeuvre', follows that here again, – just like in the case of the personal sector of the third function – we are freed from the necessity of using and defining the concepts 'corporate author' and 'corporate oeuvre' alike. We need not distinguish either between authorial and non-authorial performances of bodies, or between performances that do and do not form a constituent part of a corporate 'oeuvre'. The only distinction relevant to us is between those productions and activities of a body which *are* recorded in the catalogue autonomously, and those which are *not*. This is, however, a matter which here we need not bother about, as it is already definitively settled when the elemental objects to be recorded are selected. Every elemental object recorded under the name of a corporate body becomes automatically a component of the respective composite object of the third function; and there is no possibility, either, of adding to its components any item originating from another source. So, as we have already mentioned when discussing the first function, our survey of the elemental objects of cataloguing answers also the question as to the components of the objects, and together with this also the question as to the scope and limits of the corporate sector of the third function.

Parenthetically, since from the point of view of the third function it is not only unnecessary, but also impermissible to distinguish between the authorial and non-authorial performances of bodies, it is obviously impossible to derive from *this* function any arguments against the discarding of corporate main entries.

72. Overlapping objects in the corporate sector of the third function. We have to deal this time again, as we had to also in the case of the personal sector, with a question of detail, which is not answered by our reference to the fact that the aggregate of the components of the objects of the third function is determined by the aggregate of the elemental objects recorded. There are a few special cases in which two related composite objects of this sector overlap, in consequence of which it may become necessary to record a single elemental object in the capacity of a component of two or more objects of this sector. All these cases must be considered separately, because the objects in each of the dubious areas call for a special treatment. This task is in its nature similar to a task we had to deal with when discussing the personal sector; the concrete problems arising here are, however, of an essentially different nature. This is another proof of the fact that be-

tween personal and corporate authorship there is very little relationship indeed, not only theoretically, but also from the point of view of cataloguing practice.

The difficulties here referred to arise with three different kinds of elemental objects not entered under the name of a personal author. Firstly, with those in the production of which several different bodies have had a part; secondly, with those which bear along with the name of a superior body also the name(s) of one or more subordinate bodies; and thirdly, with those published by a body whose name has undergone a change. We have to answer the following question: under which of the more than one corporate names borne by it, or in some special way connected with it, should an elemental object belonging to one of the above three groups be made accessible by the catalogue; the question of the form of entry (main entry, added entry, or reference) to be used for attaining this end remaining here just as irrelevant to us as it has been before.

With the problems of cataloguing the items belonging to the first group, and with the shortcomings of the codification governing this field, we have already dealt with in our survey of the elemental objects (cf. p. 85); so now we need not take up this topic any more. As regards the second and third groups, international cataloguing practice has already contrived to find solutions which seem satisfactorily to settle their treatment. Let us inspect these solutions.

The members of the second group bear along with the name of a superior body also the name of one or more subordinate bodies which are no simple departments or organs of the former, and have also names of their own. As to the cataloguing of an elemental object belonging to this group, the great majority of cataloguers seem to agree that it is necessary to refer from each body under the name of which neither a main nor an added entry has been made for this object, to the body under the name of which one of these entries has been made. This is equivalent to making the object in question accessible in the capacity as a production of each body named on it, and so a component of all the third function's individual objects concerned.

The most problematical among the three groups of elemental objects is the third one, since a change of name may be connected with a substantial change in the body itself, i.e., with the extinction of the old and the formation of a new body. The cataloguer is in such a case faced with the question, whether it is two different bodies, and two different objects of the third function, that he has to deal with, or only a single one. This question often cannot be answered unequivocally. There we have again an essential difference between the personal author and the so-called corporate author. With persons a change of name generally does not raise any doubt as to the identity of the bearer of the two names; with corporate bodies it very often does. This difference – together with that other one that in the case of bodies it happens much more frequently than with persons that in his efforts to deal appropriately with a change of name the cataloguer is thwarted by a lack of information – justifies with regard to bodies which have borne successively different names, a treatment different from that generally applied to persons who have changed their name. Weighty arguments might be raised even against the collocation under a uniform name of all the entries relating to such a person; but the arguments in favour of pooling in this way all the information about the same person remain still conclusive enough. In the case of altered corporate names the two above-mentioned factors reverse the balance of advantages and drawbacks: it is certainly more expedient, chiefly because less confusing for the readers, to deal with all such cases consistently as though one had to deal with two (or several) different bodies. Recently this view seems to have been spreading rapidly; most modern codes and drafts – the *Lubetzky Draft*, the *Paris Statement*, the *West-German Draft*, the *Anglo-American Cataloging Rules*, etc. – agree in altering the traditional treatment of bodies with two or more different successive names, by switching over to treating them, even when the change of name has not been connected with any substantial change of the body itself, as though they were two (or more) different bodies.

This solution has, of course, also a drawback: it dismembers the production of many a body constituting no less a real and solid unity than those of bodies which have never changed their name. From the point of view of logic it is, thus, rather an arbitrary solution. An indispensable corrective of it is, therefore, a system of cross-references connecting *all* successive names: those belonging to an identical body as well as those known to have belonged to essentially different bodies superseding one another. This device renders all the information about a substantially unchanged body, as well as that concerning several different bodies superseding each other, retrievable under every successive name; though, of course, partly indirectly, by means of references, in two or more moves. If the change of name has been connected in fact with a substantial change, with a loss of the body's indentity, this solution undeniably transgresses the boundaries set to the third function by logic – but only in a way which does not impair the efficiency of the catalogue.

73. THE THIRD FUNCTION IS A SEPARATE SELF-CONTAINED TASK. We come now to our last task: to answer the three questions, already familiar to us, concerning the standing of the functions within the hierarchy of the tasks of the catalogue; this time with regard to the third function.

We propose to begin, here again, with an examination of the second question: whether it is right to consider the task we have been calling the third function as a separate function distinct from the other two functions of the catalogue. Our reversing the logical sequence of the three questions is this time warranted by the fact that the third function is not usually represented as a separate function, but only as a part of a comprehensive function the other part of which is what I have designated as the second function and represented as another entirely separate, in itself complete function. By this still very popular conception the number of the functions is thus reduced to two. Since most cataloguers are accustomed to look at the third function in the light of this conception, it seems advisable to begin our examination of its status with a scrutiny of this point. Besides, a certain intrinsic relationship between these two functions, already touched upon above, is indeed apt to give the impression that these are not in fact two separate functions, but merely the two integral parts of a single one. After all, the essence of both consists in collecting and appropriately arranging material which is inwardly, intellectually, and not formally, materially related; while the means, the devices used in performing them are in principle analogous, and in practice even partly identical.

Before starting to examine this problem, we had better answer the question, whether or not the third function is a function separate and distinct from the first one. This may be done without any difficulty. Between the first and the third function there exists no inner relationship of the kind which connects the second and the third with each other. The first function does not involve any pooling or arranging of information on several related objects; as we have seen, its essence consists in recording the elemental objects as entirely separate and detached items, as the units of which only an incoherent, unrelated, atomized mass is composed. Moreover, the two functions are not only essentially different in their character, but in some respects even antagonistic to each other; this is clearly shown by the fact that weighty arguments can be advanced in favour of assigning the main entry to either of them. There cannot be any doubt, then, about their being two distinctly separate functions.

The problem of the relation of the third function to the second is less easy to deal with. The two functions are, I repeat, usually represented as constituting a single function. Even the *Paris Statement* adopts this conception; it talks explicitly of the *two* functions of the catalogue.[55] Cutter seems to have discerned that the second and third functions are entirely separate and distinct; his not quite adequate definition of the second function, however, did not express this lucidly enough (cf. p. 49), and so his opinion remained unnoticed and sank into oblivion. The first author to resume, after more than three quarters of a century, and to give an articulate expression to Cutter's hint was Eva Verona, who made a point of distinguishing three different functions of the catalogue, and emphasized that to communicate information about literary units is a function distinctly separate from the function of providing information about authors' 'oeuvres'.[56] This was so obvious an improvement on the generally accepted standpoint that since the appearance of her paper several authors have already dropped the old formula and adopted her conception without acknowledging her priority, and perhaps even without realizing that by advancing it she has eliminated a long ingrained slip of logic from cataloguing theory. In what follows I wish to round off E. Verona's very apt argument on this question.

The still prevailing mode of representing the second and third functions is, I think, rooted in reminiscences of an earlier stage in the development of the author and title catalogue. I mentioned that, by using the conventional forms of authors' names, this catalogue had already managed in its infancy not only to perform the third function, but also to come rather near to fulfilling the second. The fact that the most important device applied in performing them (the use of a uniform author's name as heading) is common with both these functions, might have led to their becoming commingled in librarians' ideas. Later, when cataloguers began to acknowledge, on the one hand, the principle that the entries of the different edi-

[55] *Statement of Principles*, § 5 (Cf. our Note 13).

[56] VERONA, E.: Literary Unit Versus Bibliographical Unit. *Libri*. Vol. 9, No. 2, 1959, p. 79. – Lubetzky did not fail to recognize the validity of this proposition of Verona. In his *Principles of Cataloging* (see our Note 33) he maintains that "The treatment of a publication as the representation of a particular work by a particular author implies three levels of identification – of the publication itself, of the work represented by it, and of the author of the work" (pp. 100–101), and points out that the bringing together of all the books of an author under a particular name is, and formerly in fact was, quite compatible with a neglect of the collocation of all the editions of his particular works, i. e., with a neglect of the objective of conveying adequately arranged information also about single works. In spite of this, curiously enough, he does not jettison the traditional formula according to which the catalogue has only two, and not three, functions (see p. 14 of the same publication).

tions of *anonymous works* should also be brought together, and on the other, the necessity for a subarrangement by literary units of the entries made under the name of a particular author, the new devices they had to apply for these new purposes must have made them realize that the task we call the second function is a task different from the one we call the third function. The conflict of this theoretical discovery, made by the advanced few, with the notions of the great majority and its habitual attitude of clinging to the old-established patterns of thinking, may have given birth, through a compromise, to the formula surviving even today.

From the point of view of logic this formula is, of course, quite untenable. It implies that the two tasks under consideration constitute the integral parts of a single task, of a unity. Such a unity, however, cannot be discerned, from whatever standpoint one might try to look at the matter. The two proposed tasks cannot in fact be integrated into a single whole; they persist to remain unalterably two distinct tasks with different essences and diverging objectives of their own. There are two possibilities conceivable of the two tasks forming really one single function: if either of the two formed one of the parts of the other, or if either were one among the means of the performance of the other. Neither of these conditions is fulfilled. Of course, the formula representing the two functions as one could not be saved even by their being fulfilled, since the naming of a whole and of one of its parts, or of an end and of one among the means of its attainment, side by side as equivalent factors in a definition by enumeration, is a cogent reason for rejecting this definition.

An explanation of the amazing power of resistance shown by the tenacious survival of that self-contradictory formula might be found, perhaps, in a simple misunderstanding. In the minds of the few librarians who ever took the trouble to bother about the theoretical question of the functions of cataloguing, the notion might have been lurking, never closely examined, that the first proposition in the erroneous formula for the so-called second function referred to the task of providing information about literary units having an author, while the second to the task of communicating information about anonymous literary units. On this ground the formula could be considered as dealing only with a single function, namely, with the function of conveying information about all literary units; while its time-honoured form which splits up this single task into two parts might be regarded as justified by considerations of expediency, on the basis that in consequence of the actual duality of the means employed in performing this single task (uniform author's name on the one hand, and uniform title on the other) such a bifurcating form of the definition is more in harmony with the actual experience of catalogue-workers, and therefore more instructive and more readily accessible to them than an undivided one would be.

Such a notion, if in fact held, would, of course, be absolutely false. The difference between the two tasks under discussion is very far from amounting only to their having for objects authors' works and anonymous works respectively. Assembling all information concerning a particular person or body is a task at the same time more and also less comprehensive than the task of assembling all information about a particular literary unit. It is *more* comprehensive – as has been pointed out so ingeniously by E. Verona – because it is quite feasible to convey information satisfactorily about all the available editions of every single work of an author without collocating in the catalogue the whole available production of this author, for instance, by collecting the editions of each particular work of his under the physical marks borne by the first or by the majority of its editions. And it is *less* comprehensive, because the task of assembling all the information about the performances of a person or body is already solved if only all the entries relating to this person or body have a uniform name for heading; whereas the task of assembling literary units is not adequately performed unless the entries under this uniform name are also subarranged by literary units.

Anyhow, irrespective of the validity of our assumptions on its origin and on the causes of its persistence, the conception under consideration – acknowledging the separateness and distinctness of the second and third functions, but pretending in spite of this that the two are only one – proves either that by their reluctance to abandon an ingrained notion cataloguers are more than normally handicapped in "discovering the obvious", or else, that they do not much bother about clarifying their aims or about expressing their thoughts accurately.

Some readers, perhaps, might consider my making such a fuss about an unimportant matter of this sort, about a "mere" matter of logic having no practical significance at all, as unwarranted, or even as mere hair-splitting. If so, I should answer by stressing my conviction that logic is, even for cataloguers, no luxury; that its neglect results inevitably not only in muddled thinking, but also in inconsistent, vague practices. As an illustration, I give here a passage from a book of a pre-eminent cataloguing expert, P.S. Dunkin, published in 1969:[57]

"Then we examined the heading, the form the name of the author would take at the head of the author entry. We found that this heading must be a uniform heading for each author if entries for all the works of that author and all the editions of that author's works were to be kept together; this is the second objective."

This is to a certain extent misleading statement. It implies that by using uniform author headings, at the same time also a keeping together of all the editions of the single works by these authors can be reached. The only possible explanation for this slip, I can think of, is that, when using the expression 'second function', in his firm adherence to the old-established formula, Mr. Dunkin had alternately once only one, another time two different functions in his mind.

To sum up: one must regard the service which we have called the third function of the catalogue, as a service distinctly separate and independent from the equally distinct and independent service which we have designated as its second function. The third function is not subordinate to, nor a part of, any other task performed by the catalogue; it is a task absolutely in its own right.

74. THE THIRD FUNCTION IS RIGHTLY RANKED WITH THE 'FUNCTIONS'. We must turn now to the first of our already well-known three questions, to the question whether or not it is justified to rank this task with the functions.

It cannot be denied that by performing this task the author and title catalogue answers a demand raised by its users not only very frequently, but also very urgently. It is not only that readers expect to be provided with this class of information because they are accustomed of old to getting it, the catalogue having since its very beginnings never ceased to convey it rather adequately. But, apart from habit, there is also a more weighty justification for the users' – the readers' as

[57] DUNKIN, P. S.: *Cataloging U.S.A.* Chicago, A. L. A. 1969. SBN 8389 0071 2. Cf. p. 34.

well as the librarians' – demand for this class of information. The production of a person, and even that of a body, is a more or less organic unity, and there are many reasons for which a reader may wish to be made acquainted with the whole of this unity. True, his demand to be furnished with this information is very often fulfilled only imperfectly by the author-title catalogue, since the majority of the world's personal and corporate productions are contained only fragmentarily even in the largest libraries. Nevertheless, in the really important cases: in those of the most frequently demanded productions, the productions of universally, locally or professionally outstanding persons or bodies, the information contained in the catalogues of major research and of large libraries is usually practically complete. Librarians are, obviously, no less in need of the information provided by the third function than readers. The acquisitions librarian needs to be supported in his activities by appropriately arranged information about all the productions of a particular person or body which are possessed by his library; while the cataloguer must be given quick and easy access to all the earlier entries made for a particular person or body. (The reference librarian's needs are similar to those of the reader's, and have been taken into consideration already in what we said about the readers' needs.)

Satisfying all these needs is one of the most important services of the catalogue; a service which involves a considerable enlargement of its informational potential; and, besides, in conjunction with the second function, also a thorough transformation of its whole structure. Apart from the first two functions, there is, doubtlessly, no other service performed by the author and title catalogue of our days equal to this one in importance; and I am not sure even if the users do not derive altogether more profit from this service than from the second function. Therefore, no one who agrees that the second function has to be classified as a function (in the sense defined in Chapter II) can deny the same rank to this service. We insisted that no task should be classified as a function which was subordinate to any other task of the catalogue *and* which was not of fundamental importance. In the case of the service which I call the third function both these conditions are fulfilled. Classifications have as a rule their weak points, but the three functions we have dealt with in this study surpass so greatly in importance all the other services performed by the author-title catalogue, that the classification which exalts them, and only them, above all the other services of this catalogue has a quite exceptionally solid foundation and is uncommonly safe against objections.

75. ADDITIONAL SERVICES OBTAINED BY THE PERFORMANCE OF THE THIRD FUNCTION. It is not for the sake of confirming our proposition on its function-status, but only in order to complete the foregoing survey of the services it includes, that we must point out that by fulfilling the third function valuable services can be obtained even beyond the boundaries of its proper sphere.

The device employed in performing the third function – the collocation under a uniform heading of all entries referring to a particular person or body – is specific only to this function. All the advantages the application of this device may yield in the fields of the other two functions must be considered, therefore, accessory products of the fulfilment of the third function.

The first of these accessory services consists in helping those catalogue-users who are looking for a particular book or work connected with a given person or body, and are acquainted only with the name of this person or body, but not with the title of the book or work they are seeking. This gap in their knowledge is expediently bridged over by the collocation under a uniform name of all entries referring to a particular person or body (together, naturally, with the measures taken for directing the user to that name). This means that the performance of the third function increases also the efficacy of the catalogue in fulfilling its two other functions.

Another accessory service of the third function consists in supplying the cataloguer with the device best suited – except, of course, in the case of anonymous works – to serve the performance of the second function. As pointed out by E. Verona, it is possible to perform this function without collocating under a uniform name the entries referring to all the productions of a particular person or body, by means of assembling the information about each particular production under a person's or body's diverse names which is borne by the first, or by the majority of the editions of that production. Yet, it is evident that both these methods would cause readers as well as librarians much annoyance, since both are liable to locate the information about many literary units under unfamiliar names, or forms of name of authors or bodies. This would be not only cumbersome, but also likely sometimes to become directly misleading. With regard to non-anonymous literary units, the most adequate way of fulfilling the second function is, therefore, collocation under persons' and bodies' uniform names, i. e., the application of the device specific to the third function.

76. THE PLACE OF THE THIRD FUNCTION IN THE RANKING ORDER OF THE FUNCTIONS. This question, just like the corresponding ones concerning the other two functions, cannot be answered unequivocally. If, however, we stick to the point of view which has induced us to accept the current opinion on the priority of the task usually called the first function, then we must conclude without doubt that the ranking order of the second and third functions we have provisionally established in Chapter II must be the appropriate one. It is the logic of systematization which compels us to accept this order, seeing that the second function is more closely related to the first function than the third. It is owing to the fundamentally formal character of the author-title catalogue of our days, that we have come to consider the first function, in which that formal character is most clearly asserting itself, the most markedly intrinsic task of this catalogue. The third function, on the other hand, has, as we have pointed out, a rather pronounced systematizing character which is to some extent directly in contrast to the formal nature of this catalogue. The second function has its place in the middle between these two. It involves, as distinct from the first function, but only in a lesser degree than the third one, the rearrangement on a non-formal basis of a considerable part of the information communicated by the catalogue. Its objects are, again in contrast to the first function, composite objects, but they are composed of more homogeneous elements than those of the third function. An object of the first function is for the cataloguer a simple, single unit, from the material point of view as well as intellectually, i. e., from the point of view of its content element. An object of the second function is

for him such a unit only if seen from the side of the content element, i. e., only intellectually while materially it is already a composite, a complex object. Whereas the objects of the third function have in both these respects a composite, complex character, and form units only in a third respect,but even looser units than the objects of the second function represent.

The second function is, thus, in character altogether less removed from the first function than the third; and so it is to this second function that we must assign the rank next to the first function. The secular development of the author and title catalogue, and principally the fact that in the course of this development its character has become definitively formal, has, then, reversed the position of the first and third functions: from being practically the first, the third function has been turned by these factors into the logically last, and the first from the last into the foremost function of the catalogue.

The precedence of the second function over the third based in the above argument on logical grounds, is in itself, naturally, not a sufficient reason for granting the former also practically, i. e., in our catalogues, a more prominent position than to the latter. Such a policy could not be vindicated except for adequate practical reasons. But the reverse holds true, too. The policy which allots to the third function the benefit of being fulfilled on the plane of the main entry, but refuses to accord the same benefit also to the second function, calls no less imperatively for a justification by adequate practical reasons. However, so far as I know, the numerous advocates of such a policy have as yet not been capable of offering such reasons.

*

I have come to the end of my inquiry: Of course, I do not assume that I have contrived completely and definitively to settle all the problems which I had raised. But I do hope that I have thrown at least some light upon hitherto neglected questions of cataloguing theory, by pointing out one or the other significant connection, the proper place of some relevant facts and factors within its system. The mistakes which I have surely committed and the problems I have overlooked, will certainly not remain unnoticed, and their exposure will, I hope, help to promote the very case which I too wished to serve: the advancement of cataloguing theory.

It is for the same reason that I also cherish another hope, the hope that I did not fail to reach my second aim, the verification of my two premises stated in the preface: first, that in author-title cataloguing present-day views and conceptions are still often logically incorrect, generally accepted definitions superficial, terminology inaccurate, and second, that this inadequacy of cataloguing theory in fact results in mistaken practices and even in some mistaken objectives. If I have succeeded in demonstrating these two points convincingly enough, then I have also contrived to perform a not valueless positive service: to show that cataloguers ought to care much more for the theoretical foundation of their art than they actually do.

APPENDIX I

HAS IT NOT BEEN UNWARRANTABLE TO RESTRICT THE SCOPE OF THIS INQUIRY TO STANDARD CATALOGUING ONLY?

1. SHOULD ALL THE BASIC RULES OF CODES EQUALLY APPLY TO ALL LIBRARY MATERIALS? We raised the question in the title of this appendix already in Chapter I, and mentioned there provisionally that our answer to it was negative, but postponed the justification of this assertion. The object of this appendix is to supply that justification.

It has been variously maintained that, so far as the choice of the heading and entry word is concerned, the proper way of composing rules for author-title cataloguing is to make them valid uniformly in all sectors of this branch of cataloguing. This position implies a condemnation of the traditional usage of having apart from the rules for standard cataloguing also separate and different bodies of rules for each particular field of special cataloguing. The logical inference from this view is that cataloguing theory must be one and uniform too, i.e., it must equally apply to all sectors of author-title cataloguing. This inference, if valid, implies also a rejection of the limitation of our subject which this appendix is concerned with. Such a limitation would, obviously, not make any sense unless we assumed that cataloguing theory, or at least its chapter on the functions and objects of cataloguing, ought to be elaborated separately for each sector of author-title cataloguing.

Let us begin our examination of this argument with an analysis of the thesis that the rules for the choice of headings and entry words must be applied uniformly in all sectors of author-title cataloguing, inclusively of the cataloguing of materials produced by technologies other than printing.

Initially we must emphasize that the demand for an integrated system, with closely co-ordinated rules for the different kinds of library material is, obviously, absolutely justified. But this demand seems to get easily mixed up with that other, unjustified one, that the cataloguing of all kinds of material should obey the selfsame fundamental rules. To avoid misunderstandings, it should be emphasized that we are not concerned here with the question whether the European practice of establishing special catalogues for special materials is right or wrong, but only with the entirely different one as to whether the present world-wide practice of applying to the diverse special materials different, specific rules, among them also fundamental ones, is objectionable or not?

The standpoint that the rules concerning the choice of the heading and entry-word ought to be applied uniformly in all sectors of cataloguing is erroneous. I propose to demonstrate this by analyzing a particular practical experiment of realizing it, incorporated in § 1 of the *Lubetzky Draft* of March 1960. Before doing so, I wish, however, to emphasize parenthetically that, as most of us, I too regard this Draft as the most outstanding achievement known in cataloguing codification up to the present day.

The paragraph mentioned ventures to put forward a fundamental rule the validity of which is not confined to printed documents only. The fact that this proposal has been dropped in the meantime, and the question whether this was done for essential reasons, for reasons of logic, or only on practical considerations, on considerations of expediency, are of no relevance to us, since the sole reason for the following lengthy analysis of the proposed rule is its extraordinary aptitude to serve as an illustration of the erroneousness of the proposition which we are at present engaged in examining.

The rule in question runs as follows: "The work of a person, whatever its character or the medium by which it is presented, is entered under the name of the person as author of the work presented". The rule is followed by a note intended to define the concept 'author': "The author is the person whose work is presented or exhibited, not the one who prepared a particular edition or presentation of the work. When the art or skill of a performer is exhibited, the performer is treated as the author of the work exhibited". The examples attached to the rule refer partly to books, partly to gramophone records. This indicates the kind of performers to whom the second sentence of the note alludes.[58]

The first sentence of the note is not at variance with the meaning the word 'author' bears in everyday speech. This is, however, its only merit, since as a definition of the concept 'author' it is, in consequence of its vagueness, of no use. The purport of the second sentence, on the other hand, does not conform at all with the current notions about authorship, and what is more, it flatly contradicts the first sentence. The art or skill of the performer is, so far as gramophone records are concerned, without exception always "presented or exhibited". So, according to the second sentence gramophone records ought to be entered without exception under the name of the performer, whereas according to the first they should never be treated in this way. Among the two sentences, it is evidently the second which is not in accord with what the authors of the Draft really meant. Their actual intention might have been to have records entered under the name of the performer when he is considered more important or interesting than the author. If so, their intention was absolutely false; such an interpretation could not save their endeavour: it is obvious that a rule involving the necessity of comparing the importance of the composer (or author) with that of the performer – without exception, of course, as the rule gives no qualification – is inevitably destined to fail. Such a rule would be workable only in extreme cases, when the performer is undisputedly first-rate, whilst the composer (or author) only third-rate, or conversely; but in the majority of cases it would prove simply meaningless, as it would be based on an absurd criterion.

Among the examples attached to the rule cited, there are only two which are relevant to the problem under consideration. Since both of these relate to

[58] *Code of Cataloging Rules. Author and Title Entry*. An unifinished draft prepared by S. Lubetzky. Ed. by the A. L. A. March 1960. § 1.

gramophone records containing several pieces of music not composed by the same person, but performed by the same artist, we had better emphasize perhaps that the creators of this unconventionally comprehensive rule cannot possibly have had the intention to restrict its validity to this special sort of records only, and to secure for these a treatment analogous to the one widely applied to anthologies in book form. To introduce such a petty matter into the code's very first rule, devoted to a fundamental matter of in the highest degree general importance, would have been a quite unpardonable lapse of logic, which we must not impute to the authors of the rule.

At any rate, the mistake the rule in question displays is a glaring one. The only possibility of explaining it seems to be that it must have originated, during the debates of a committee, in a momentary prevalence of the supporters of the view that it was desirable to have fundamental rules with the most possibly universal validity. That the rule stops dead half-way, and fails to take into consideration the specific circumstances one has to deal with when cataloguing special library materials other than gramophone records (e.g. motion pictures), is another, though much less heavy, logical shortcoming of the draft rule.

2. SHOULD ALL THE BASIC RULES ALSO APPLY TO DOCUMENTS NOT PRINTED? The important thing from our point of view is not that this particular attempt at formulating a fundamental rule valid also for materials other than printed documents has failed. It has been carried out with such a glaring cursoriness which might seem to account for the failure sufficiently, and give rise to the thought that with more care it could have been made a success. The really important thing is that even the utmost care would not have made it possible to achieve this. The attempt was in advance condemned to failure, in consequence of its violating the basic principles governing the composition of cataloguing rules.

Perhaps the most difficult of the tasks the compilers of codes have constantly to cope with is to strike the usually very narrow happy middle course between two opposite mistakes. The first of these consists in making special rules, divergent from the pertinent general one, for rare specific cataloguing conditions which can be dealt with without any serious drawbacks according to the general rule; calling at the most for additional secondary entries. Superfluous provisions of this sort gratuitously impair the simplicity and clarity of the structure of the author-title catalogue, and increase the difficulties which readers have to surmount when consulting it. And what is still worse, by the creation of new border-lines separating the adjacent spheres of application of possibly concurring rules, they give the reader as well as the cataloguer a good deal of additional trouble and annoyance. For the rule-maker it is only exceptionally possible to fix with precision and clarity the position of such a line of division, and to avoid the creation of an unsafe borderland; which is in its turn bound to become a prolific breeder of cataloguers' mistakes and inconsistencies. The quantity of such mistakes and inconsistencies increases, thus, in proportion to the total length of the border-lines the establishment of which is implied in the rules of a code.

The opposite mistake which authors of codes must be careful to avoid, is the subsuming of essentially different cataloguing conditions under the same rule. Rules of this sort are liable – due to the incongruity of the different sorts of objects they should cover, and to the use of vague or equivocal concepts and terms to which they, in consequence, are often compelled to resort – to turn out obscure, and so the limits of their sphere of application are bound to be more than usually liquid. Obviously, the detrimental results of the first mistake may be tempered or even eliminated, if it is possible to draw a sharp line of division, based on clear criteria, between the two related situations with which the two rules are dealing. Conversely, the latter mistake is rendered by the same possibility quite inexcusable.

The draft rule we are at present engaged in analyzing is, obviously, a characteristic instance of the latter mistake. The conditions one has to deal with when cataloguing printed documents and gramophone records, respectively, are, first of all in consequence of the emergence in the latter case of the new cataloguing object "performer's performance", fundamentally different; and this new object involves even the appearance of a function unknown to standard cataloguing. Subsuming matters in essential respects so heterogeneous under the same rule, is a grave mistake. The more so, as the chief argument against a splitting up of one rule into two separate and concurrent rules – the lack of clear criteria marking with exactness the line of division between the two groups of cases to be covered by the two rules – does not apply here at all: the striking technical dissimilarity between books and gramophone records, and the no less obvious difference between author and performer provide the necessary criteria in abundance.

The foregoing analysis of § 1 of the *Lubetzky Draft* has shown not only that this paragraph did not succeed in escaping the Scylla and Charybdis of code-composition we have just pointed at, but also that any other attempt at achieving what it did fail to achieve, would be no less liable to fail. This implies that the demand for uniform fundamental rules, to be applied to standard cataloguing as well as to the cataloguing of library materials produced by technologies other than printing, is a mistaken one.

3. SHOULD ALL THE BASIC RULES ALSO APPLY TO PRINTED DOCUMENTS NOT TO BE SUBJECTED TO STANDARD CATALOGUING? Everything we have said about the documents produced by technologies other than printing applies also to the two genera of printed documents that do not belong to the domain of standard cataloguing. (It should be recalled, that in this context the term 'printed documents' is used with a denotation including also the documents produced by technologies related to printing, e.g., by mimography.)

The first of these two genera of documents is constituted, as already mentioned, by certain special classes of printed documents using means of expression other than written language: by printed maps, music, and pictorial representations. To the rules providing for the cataloguing of documents of the first two classes most codes devote separate chapters or appendices; whereas to the documents belonging to the last class, if they have the outward appearance of books, the rules for standard cataloguing are usually applied. In our days, I think, the latter practice has already become a slip in logic, a survival from the heroic age of cataloguing, displaying the style of this age, and lagging as to articulateness, lucidity and precision far behind the present-day standards of cataloguing codification. As a matter of fact, in the course of the cataloguing of documents belonging to any of the three classes there may arise the necessity for an autonomous recording of specific agents participating in their production, i.e., the necessity for a recording of autonomous objects alien to standard cataloguing; and this might sometimes even generate specific functions.

The second of the above-mentioned two genera consists of such printed documents to which, for different reasons, a greater or a lesser significance is attributed than to the great bulk of printed documents, and which are catalogued, in consequence, with either a greater or a lesser minuteness than the one applied to standard cataloguing. In the former case there may ensue the appearance of objects and functions unknown to standard cataloguing; in the latter the discarding of some objects or even functions of standard cataloguing.

To some species of documents belonging to this genus a special treatment is applied rather generally and consistently, e.g., to incunabula at the upper end of the scale, and to ephemera at the lower end. Other kinds have a far less stable status. For instance, some national libraries accord to 16–17th century books printed in their home country a considerably more detailed treatment than the standard one: a so-called bibliographical treatment. On the other hand, some municipal and regional libraries apply to the ephemera and fugitive material of local concern a standard treatment instead of the perfunctory brief-listing they apply to the non-local items belonging to these categories. The catalogical status the majority of libraries grant to the same documents is in both cases one degree lower.

Here we have come across such kinds of documents as have an unstable catalogical status, similar to that of the objects which are not always, only potentially, autonomous. The existence of these kinds of documents does not entail any corresponding vagueness of the limits of our subject-matter. In this respect, these limits are not bound up with specific kinds of documents, but with a particular method of cataloguing: with standard cataloguing. If a document is catalogued in different libraries alternately according to the rules of standard and bibliographical, or of standard and simplified cataloguing, it is irrelevant to our subject that in one of these cases this document is not an object of standard cataloguing: in spite of this it belongs, owing to the other case, unquestionably within the scope of this inquiry. So the existence of these sorts of 'amphibious' documents does not in any way increase the intricacy of the limits of our inquiry. All of these documents do form a part of the subject we have to examine, as they *are* being treated, albeit inconsistently, according to standard methods. We may also mention parenthetically that it is no business of ours to worry about the location of the boundary line between the spheres in which different methods of cataloguing are to be applied to these kinds of documents. This is a matter not to be decided by cataloguing, or any other, theory. It is rather one of practical policy, which each library must settle for itself, according to its peculiar objectives.

4. THE HABITUAL STRUCTURE OF CODES IS A HAPPY ONE. We may, then, confidently declare that the traditional and habitual mode of systematizing cataloguing rules, i.e., the custom of splitting up the whole body of these rules into separate codes, or into sections or chapters of a code according to (*i*) the technology employed in the production of the documents catalogued, (*ii*) the means of expression used in them, and (*iii*) their comparative significance – is fortunate and advantageous. It is simply impossible to shape all fundamental rules so as to give them uniform validity in all sectors of author-title cataloguing.

This does not imply, however, that it is either needful or expedient to prepare a complete separate code for each type of library material. Analogous situations call clearly for uniform rules, even if they happen to occur in different sectors of cataloguing. It is, for instance, obvious that the choice of uniform personal names must be uniform in all sectors. In other words, in spite of the fact that the domain of cataloguing is split up into several sectors most of which are in need of some separate fundamental rules, the situations and problems arising in several or in all sectors in identical form, validate once more the first of the two antithetical principles of code-making earlier indicated (cf. p. 138.). Traditional practice is, thus, right in this respect too: it is sufficient to prepare a complete code only for a single sector of cataloguing,whereas to the bodies of specific instructions which all the other sectors need it is more expedient to give the form of mere appendices or supplements to the former, to the fundamental code. These supplements should of course contain only rules establishing measures and devices which deviate from those the fundamental code prescribes.

It is a matter of common knowledge that the fundamental codes at present in use are devoted to the rules of standard cataloguing, i. e., to the rules for the cataloguing of the great bulk of books and book-like materials, of the 'normal' books, the books which are allotted no specific treatment. The reasons for this are partly historical, but chiefly practical. When the first rules and codes were designed, the importance of special materials was still very slight compared with the importance of the books not belonging to any special class; and so the first rules and codes were bound to be rules and codes for the cataloguing of 'normal' books. In spite of the fact that since that time special materials have not only multiplied but also grown considerably in significance, 'normal' books have preserved their prominent position, and constitute still by far the most copious and important class of the sources of information committed to the charge of libraries. So the present custom of devoting the fundamental, the complete code to the cataloguing of this class of books and book-like materials, does not only root in tradition, but it is also still fully justified by present-day conditions. The facts that we are accustomed always to discuss the problems common to the entire domain of author-title cataloguing in relation to standard cataloguing, and that all the bodies of rules governing the cataloguing of special library materials are, on the whole, reduced to supplements to a code for standard cataloguing, are the direct and obvious consequences of the dominant status of 'normal' books and book-like materials. It is this status which also justifies the application of the attribute 'standard' to this branch of author-title cataloguing.

5. THE FUNCTIONS AND OBJECTS OF CATALOGUING ARE NOT IDENTICAL IN ALL CLASSES OF LIBRARY MATERIALS. We have now got an answer to the question preliminary to our original problem. We have found that the present-day general custom to prepare, not a single universally valid code, but along with a fundamental code for standard cataloguing, also code-supplements for each special class of library material,is not only unobjectionable, but really the most appropriate and expedient policy. Now we may pass on to the question which is our real concern: whether a similarly ramified *theory* of author-title cataloguing is unobjectionable too; more concretely, whether it is logically permissible that cataloguing theory should propound propositions about fundamental problems which are not equally valid for all classes of library materials.

This question cannot be answered unequivocally. The initial chapters of the theory of author-title cataloguing which, obviously, must deal with the objectives and the scope of this theory itself, are likely to contain mostly generally valid propositions. But in the chapter dealing with the functions and objects of cataloguing, to which this inquiry is devoted, such a unity and universal validity cannot be attained any more. From our findings concerning the best way of organizing codes, we must infer that in this of its chapters theory, too, must deal separately with each sector of author-title cataloguing, with each particular type of library material. We have seen that in most special sectors the cataloguer is obliged to record peculiar objects which he never meets either in other special sectors, or in the domain of standard cataloguing, and that, conversely, there are objects familiar to standard cataloguing which never occur in one or more special sectors. And since the number and scope of the functions the catalogue actually performs depend on the range of the objects it records, the functions in special areas of cataloguing are also likely not to conform entirely with those performed by standard cataloguing. Therefore, theory must scrutinize the functions and objects of author-title cataloguing separately for each particular sector, and so its chapter on them must be ramified. In the subsequent chapters of theory the same need to change over continually from unified to ramified treatment and back, very probably persists. So, for instance, the choice of the appropriate form of a particular heading (e. g., of a certain sort of author heading) must be taken uniformly in all sectors of cataloguing; and so the question of the choice of the form of headings calls in theory, too, for a uniform and general treatment. On the other hand, information on the technological peculiarities of each particular class of library material can only be dealt with separately, in codes as well as in theory.

Now, at last, we have reached the stage of being able to answer the question raised in this appendix. Since theory must investigate the functions and objects of author-title cataloguing separately for each of its sectors, and since by far the most important among these sectors is standard cataloguing, our having confined this inquiry to the latter is logically unobjectionable. On the contrary, with the help of this simple device we have succeeded in securing the double-barrelled advantage of embracing all facts and problems of general concern, and of excluding at the same time most of the comparatively unimportant and uninteresting details.

6. THE SPECIFIC PROBLEMS OF PUBLIC LIBRARY CATALOGUING AND THE TOPIC OF THE CATALOGUING OF PERIODICALS. Standard cataloguing has been defined above as a branch of cataloguing governed by the rules which virtually constitute the contents of traditional national codes designed with a view to the needs of large general and research libraries. In smaller libraries and in most public libraries cataloguing differs in many respects from standard cataloguing. In some countries, therefore, there are prepared – very appropriately – special codes for the use of the latter classes of libraries.

Most of the differences betweeen standard and public library cataloguing consist in the latter's neglect of objects the recording of which is considered indispensable by the former. These differences are, of course, of no interest whatever for our subject: its being dropped by public library cataloguing does not lessen the importance of a function or object for us. But there is also a surplus in public library cataloguing – an outcome of the chief function of this class of libraries, of their being among the foremost agents of popular education. This function of theirs, alien to research libraries and to most big general libraries, gives also their author-title catalogues a new function: to disseminate elementary knowledge, e. g., by means of annotations attached to their entries. This specific function of public library catalogues is the third topic we exclude from our inquiry. We do that not because we consider this function inferior, but merely since it is not an intrinsic task of author-title cataloguing, but only secondarily imposed upon it in special circumstances; and also because, as a rule, it is not productive of new autonomous objects of cataloguing, and so not relevant to our subject-matter.[59]

Finally, let us insert here also the discussion of a possible objection, questioning this time not the exclusion from, but on the contrary, the inclusion into, our inquiry of a particular topic.

From what we have said above follows that we do not propose to classify periodicals as a special library material: they fit into none of the three classes of the special materials as specified above. One might object to our having taken this position on the grounds that, due to their publication being a long-drawn process, periodicals display specific features which must not be left unrecorded, and thus call not only for entries of a special structure, but also for the creation of special rules – just like all the materials we have classified as special. In accordance with this, in many codes the special rules for periodicals are arranged so as to form separate chapters; this is a further resemblance between periodicals and the materials we have classified as special. Does not prove all this sufficiently the inconsistency of our classification, and of the logical necessity of classifying periodicals not as a standard material, but as a type of special material?

In the present context the answer must be in the negative. The peculiar features of periodicals do not entail any deviations from the functions and objects emerging in the cataloguing of non-periodical publications. Most rules in the codes' special chapters on periodicals are concerned merely with matters of descriptive cataloguing in the narrower sense of the term, in other words, with subordinate objects only. And even those few among them which do deal with the choice of the heading (e. g., the modern rule for the treatment of serials of which the title has changed) create no kinds of objects, either, which do not occur also in the cataloguing of non-periodical publications; they differ from the corresponding rules governing the latter in only one, for us entirely irrelevant, matter: in a reversal of the employment of main and added entry. So there is no reason for not subsuming periodicals along with other books under the same head, nor for not including them in the subject-matter of this inquiry.

Apart from this negative argument there is, however, also a positive one in favour of this position, including of periodicals, and a very weighty one at that. All the special materials we have disregarded are of comparatively little importance; this is the main reason why in this paper we did not wish to enlarge upon the peculiar tasks which stem from their specific features. Periodicals constitute, however, a most important and valuable class of documents, not inferior to books in any way. When discussing major issues either of cataloguing theory or of code-making cataloguers never disregard them; and it were simply absurd if they did. This applies, of course, to this study too.

[59] Cf. § 6.

APPENDIX II

CLASSIFICATION OF CATALOGUE ENTRIES ACCORDING TO THEIR OBJECTS

1. THE RELEVANCE OF THIS TOPIC TO OUR INQUIRY. In the main body of this study we have scrutinized and classified the objects of cataloguing with regard to their instrinsic nature, paying only very little attention to the correlations in existence between them and their catalogical vehicles, the devices used for making them objects of cataloguing: the catalogue entries. In Chapter V we could not avoid repeatedly to touch upon this topic, but we did so only casually. We were not in the position there to enter into its details, because this would have excessively delayed the progress of the discussion of the subject then in hand: of the specification and classification of the non-primary elemental objects of cataloguing. As a result, there have remained unclosed gaps even in the discussion of the latter subject, the main subject of that chapter. The topic of surveying and classifying the objects of cataloguing, and the topic of the correlations between these objects and their vehicles, the entries – which latter necessarily includes also classification of entries according to their objects – are very closely connected. In order to close the gaps in the argumentation of Chapter V, the latter topic must be taken up once more, and dealt with in all its details.

This is the reason for our adding this second appendix, which is, then, imperatively called for by logic. But the pursuit of the theoretical end of meeting the logical requirements originating in our intrinsically theoretical subject will yield this time too, as it has kept doing previously,also practical results. A careful classification of entries according to the objects they are correlated to, will enhance our understanding of the complexities of their services; and our better understanding of the latter will, in turn, help us in using the various kinds of entries more logically and efficiently.

For the purposes of the following analysis we propose to introduce, for brevity, the term 'object of an entry' (already applied in the title of this appendix) as a substitute for the voluminous phrase 'an object of cataloguing about which a particular entry is conveying information'.

2. ENTRIES WITH SEVERAL OBJECTS. OBJECTS THE RETRIEVAL OF WHICH IS PROMOTED BY SEVERAL ENTRIES. As pointed out before, there are two circumstances by which the correlations between the objects of cataloguing and the catalogue entries are rendered intricate and difficult to disentangle: there is perhaps not a sin-

gle entry in our author-title catalogues the function of which is confined to securing access to a single object only; and, on the other hand, the retrieval of most objects of cataloguing is promoted by more than one entry. The first circumstance will compel us to examine the ranking order of these multiple objects; in other words, before embarking upon our task: a classifying of entries according to their objects, we shall have first to classify the objects correlated to one and the same entry.

In Chapter V our interest in the theme of several objects to one entry was in essence confined to only one of its aspects: there we have been concerned merely with the principal object of entries. Now we shall have, naturally, to drop this restriction, and to extend our investigations also to the secondary, the incidental objects of entries, and to classify the latter also according to these among their objects. This does not imply, however, an extension of our investigations to the lowest and least significant class of the objects of cataloguing: to the subordinate objects. Except in a few final remarks, we shall continue to stick to the restriction imposed upon our inquiry in § 6: we shall concern ourselves with autonomous objects only, with the objects which cataloguers think important enough to devote separate entries to, entries made under these objects' formal marks.

3. PRINCIPAL AND INCIDENTAL OBJECTS OF ENTRIES. We cannot cope with the complex task of classifying entries according to their principal as well as to their incidental objects except in successive stages. We shall have to classify them first according to their principal object, this being the main determinant of their place in a classificatory system; and to deal afterwards separately with the task of classifying them according to their incidental objects. Such a *modus procedendi* obviously presupposes – just like the analysis in Chapter V did – a clear-cut definition of the *principal object of an entry*. Without such a definition we should not be able either to specify the reasons for our preparing several entries in the course of cataloguing a single book, or to find in our classificatory system the appropriate place for each among these entries.

As to the concept 'principal object', we have to supplement by some details what has already been said about it in Chapter V. We had better begin by illustrating by an example the cataloguing situation in which a single entry serves to promote the access to several different objects.

Let us take for this purpose an added entry made for the preface to a collection under its author's name. This can serve the retrieval of two distinctly different autonomous objects: of the preface itself, and of the collection as a whole. It is the reader who recalls only the name of the author of the preface, but not the title nor the name of the compiler of the collection, who will profit in the latter way from that entry. Of the two objects it is obviously the preface which must be regarded as the principal object of that entry; whereas the collection can only be considered as its secondary, its incidental object. The reason for this is that the heading of the entry – being primarily a formal mark of the preface, and not of the collection, not of the whole book – points directly at the preface, and not at the collection.

4. THE TWO STAGES IN THE IDENTIFICATION OF THE PRINCIPAL OBJECT OF AN ENTRY. The foregoing example illustrates the problem of the principal object of entries on

a rather high level of abstraction and generalization only, without taking into consideration the complications which the existence of the two cardinal classes – elemental and composite – of the objects of cataloguing involves. Let us illuminate these complications by amplifying our example by the additional information that the preface is signed by a pseudonym of its author whose original name is known.

The entry for the preface can in this case be made in two different ways: either under its author's pseudonym, i. e., under a *physical* mark of the *elemental* object concerned, or under his *uniform* name, which is a mark specific, not to this elemental object, but to a *composite* object: of that author's whole 'oeuvre'. We have now to decide whether or not we should regard and class the principal object to these two different possible entries as two different objects. The first as an elemental object: the preface in its capacity as a separate detached work; the second as an object of composite character: the preface in its capacity as a component of its author's whole 'oeuvre'.

From the point of view of logic it is impossible not to answer this question in the affirmative. Elemental and composite objects constitute in fact two opposite categories of objects of author-title cataloguing, and so the preface of our example is in its capacity as a detached, individual piece of intellectual work an object of cataloguing manifestly different from the same preface in its capacity as a particular fragment of its author's whole 'oeuvre'. The concept of the principal object of an entry as defined above includes, then, also the command to distinguish between the elemental and the composite nature of that object. In order to identify the principal object of a particular entry, one has, thus, always to perform two different tasks: first, one has to identify the individual, detached elemental object – or, in the case of references, the individual batch of elemental objects – the providing of information about which has been the main purpose of preparing that entry; and second, one has also to answer the question whether the entry provides access to this object directly, in its capacity as an elemental object (by means of its physical marks), or indirectly,in its capacity as a component of a composite object (by means of the corresponding uniform marks).

The second task is not nearly so simple as at first glance one might expect: it does not consist at all in an easy choice between only two clear-cut opposite categories. There is first of all the difference between main and added entries on the one hand, with only a single principal object, and between references, on the other, which have usually a whole batch of individual objects for their principal object. Moreover, complications of an entirely different nature originate in the manifold differences between national systems of cataloguing rules. One comes soon to realize that in this context the notion of our having to divide entries into only two cardinal classes – composed of those having an elemental object, and of those having a component of a composite object for their principal object, respectively – is a rather crude generalization which would be of very little use in the further stages of our analysis. So we have to develop a more discriminating classification, disclosing the existence of a greater number of more subtly differentiated classes.

The first source one has to turn to for an answer to both questions regarding the principal object of a particular entry, is the heading of that entry. With a majority of main and added entries, however, one cannot get an answer to the first question (concerning their principal elemental object) by consulting their heading alone, since they share that heading with the entries of several other elemental objects. In such cases one must resort, along with the heading, also to data in the body of the entry: to the title following the author heading, to the indication of the language of the translation following the uniform title of the work, to the edition statement, to the year of publication, etc.; or, in the case of added entries, to their bottom section.

5. The six main classes of the principal objects of entries. Now, even if we confine ourselves to classifying entries according to the elemental or composite nature of their principal objects merely in view of their headings, and disregard the secondary factors, the secondary data just specified, we may distinguish between as many as six distinctly separate classes of entries, which we shall specify presently. By taking into consideration also those secondary data, this number might be further increased. We do not wish, however, to enter into a specification of the secondary classes to be obtained in this way, since the instructiveness of this would not be in proportion to the trouble involved. Instead of offering such a specification, we propose only to illustrate by an example its nature and possible outcome, choosing for our example an outstanding pair of secondary classes.

The diverse alternatives of a combined application in the entry of the physical and the uniform variant of the author's name and of the title, respectively, accomplish corresponding changes in the elemental or composite character of the principal object of that entry, and in the place of the entry itself in our classification based on that character. For instance, a main entry headed by the author's uniform name which is immediately followed by the physical form of the title, has only one object of composite character: one of the components of a composite object of the third function, made accessible by the author's uniform name in the heading; but no composite object of the second function. If, on the other hand, in the main entry for the same book the author's uniform name is followed by the uniform title of the work, that main entry has two objects of composite character: one component of one composite object of each the second and the third function. This example also shows how much the lack of uniformity and consistency in international cataloguing practices does contribute to increasing the number of the entry classes of this kind.

After this digression let us return to the six main classes into which one may divide entries, merely in consideration of their heading, according to the elemental or composite nature of their principal object. Theses classes are the following:

a) Entries having for principal object a single elemental object in its very elemental capacity. This class is constituted by the main and added entries made under a physical mark of their principal object, in spite of the existence of a corresponding uniform mark. Example: a main entry for the anonymously published first edition of Daniel Defoe's *Memoirs of a Cavalier* headed by this title. Today a main entry made in this way is regarded by a majority of cataloguers as out of date, but it continues to be employed in almost all catalogues notwithstanding, though mostly on a limited scale only, principally for works of unknown authorship.

b) Entries having for principal object a single detached object, but not in its elemental capacity, but in that as a component of one or more composite objects. This class is formed by the main and added entries headed by a uniform mark which differs from the physical variant borne by the book. Example: a main entry for the first edition of *The Pickwick Papers* put under the name Charles Dickens.

c) Entries having for principal object a single object uniting in itself the character of an elemental object with that of a component of composite objects. This class is constituted by the main and added entries which have for heading such a physical mark of the object described as performs at the same time also the function of a uniform mark. It splits up into two sub-classes. The physical marks which form the headings of the entries belonging to the first sub-class have no variants at all, so that they may be regarded simultaneously also as uniform marks. The physical marks in the headings of the entries belonging to the second sub-class do not lack variants, but this circumstance is rendered irrelevant by the fact that these physical marks happen to be the very uniform marks attached to the objects in hand. Example: the main entry for a later edition of *The Pickwick Papers* published and entered under Dickens' name.

The remaining three classes of entries are composed of the diverse kinds of references.

d) The references of the first class have for their principal object one or more elemental objects. They refer from a physical mark of these objects to the corresponding uniform mark under which the main and added entries of all components of those particular composite objects are collocated which the said elemental objects form parts of. Example: the reference made in Europe from Clemens, Samuel Langhorne to Twain, Mark, and in America the other way round. A great majority of the references belonging to this class – all those which are made on account of at least two elemental objects bearing the physical mark which forms their upper part – have in a certain sense hybrid character. Since they are headed by a physical mark, their principal object is to be considered elemental in its character; on the other hand, however, the circumstance that this object is not a single detached object, but is composed of several related objects, mostly of a whole batch, seems also to confer on it a composite character. Yet, the principal objects of this sub-class of entries are as a rule composite only in a sense which sharply deviates from that we have assigned and wish to reserve for this term: they form much more often than not merely a part, a fraction, of an object of the kind to which we have proposed to apply the designation 'composite': of an object which embraces *all* the pertinent elemental objects.

e) The references – taking often the 'see also' form – which have for principal object also only a part of a composite object in the specific sense of the term, but point in a direction opposite to that which the class *d*) references point in: they direct the user of the catalogue from a uniform mark to a corresponding physical mark under which further main entries are filed. Example: Balzac, Honoré de: see also: R'Hoone, Lord. So far as personal names and single-work-titles are concerned, this class of entries – just like the entries of class *a)* – are now widely regarded as obsolete, but they continue in spite of this to survive. In a certain respect related to this class are the reciprocal references made for corporate names and serial titles that have undergone a change, the use of which is, however, in recent times steadily and vigorously spreading.

f) References having whole composite objects for their principal object, and pointing from a non-physical mark connected with that object which is not used as its uniform mark, to another which is. Example: Arouet, François-Marie: see: Voltaire.

6. THE TWO ASPECTS OF THE PRINCIPAL OBJECT OF AN ENTRY. THE 'OBJECT PROPER' AND THE 'COMPLEX OBJECT PROPER'. The last three classes in our specification may be quickly disposed of, there being very little occasion of interest for sub-dividing them. Considerably more complex than this, but at the same time also much more profitable and important, is the task of examining and classifying the single, detached, non-compound principal objects of the entries belonging to the first three classes of our specification. These objects, as well as the entries which record them, split up into numerous sub-classes, a survey of which, together with the determination of their place in our classificatory system, will constitute the bulk of the following inquiries.

Luckily, we shall not be obliged to scrutinize the three different classes separately: a single analysis will suffice to cover them all. This short cut is made possible by the fact that the most outstanding object of the entries in these three classes alike is in the last analysis a single detached, i. e., an elemental object, irrespective of the otherwise important difference that the entries in the first class record that object in its own capacity, in the capacity as an elemental object, those in the second in the capacity as a component of composite objects, while those in the third in both of these capacities simultaneously. To whichever of these three classes a particular entry may belong, the first step towards the ascertainment of its principal object must always consist in selecting the first in rank among the elemental objects the retrieval of which that entry is fit to promote; in other words, in identifying that particular elemental object by the securing of access to which that entry's main objective is formed. The question, whether the entry renders that elemental object accessible in its own, its elemental, or in its composite capacity, or in both, is, though by no means irrelevant, of only secondary importance to the classifier. It is first of all with regard to the class *b* entries that it is important to stress the precedence of the former task, their principal objects being *de facto* of composite character, as their headings are formed by uniform marks.

With regard to our first three classes of entries the identification of their principal objects is to be cut up thus into two distinct tasks, the first of which consists in an identification of the elemental aspect of these principal objects. But it is to be emphasized that the question second in rank – whether or not that individual object is presented as a component of composite objects – is still important enough. In case of a positive answer, it is indispensable to specify the composite objects of which that individual object constitutes a component. In consequence of our having to answer, thus, two distinctly separate questions, we had better examine these two aspects of the principal object of the entries under consideration separately, or even – by way of a working hypothesis – as though they dissolved (except, of course, in class *a* cases) into two distinct principal objects: an elemental and a composite one. This implies that we are in need of two different concepts of principal object and of two corresponding terms. We may define the principal object in the first sense as the elemental object a physical mark of which, or else the uniform variant of this mark, forms the heading of the entry; the testimony of the heading, if insufficient to identify unequivocally an individual elemental object, having, of course, to be supplemented by other data contained in the entry. The principal object in the second sense, on the other hand, we define as the composite object, or more precisely, a component of those composite objects, a uniform main of which forms the heading of the entry. We shall call the principal object in the first sense the 'object proper' of the entry; while the principal object in the second sense we shall designate as that entry's 'complex object proper'. We use the adjective 'complex' instead of 'composite', because we have already reserved the latter word for designating the second principal class of the objects of *cataloguing*, not of *entries*; and so its application in this second context would be liable to cause confusion.

The introduction of the concepts of two different kinds of principal objects as defined above, implies by no means that we are discarding our first, comprehensive concept of the principal object of entries. We shall continue to use it whenever there will be no reason for distinguishing between its two different aspects. It is for reasons of expediency – to be explained in detail presently – that we have been induced to substitute, for certain purposes, the two distinctive concepts for this comprehensive one. And it is for the same reasons that by our definitions of the two concepts, and by the form we have given to the corresponding terms, we have reversed the natural, the logical ranking order of the two aspects of the principal object.

For better elucidation of the implications of our two new concepts and of the meanings of the corresponding terms it will be useful, perhaps, to make here a few explanatory remarks. Class *b* entries, which form by the way one of the most common kinds of entries, have also, along with their complex object proper, an object proper. This duplicity of their principal object is given an outward expression by their being consistently accompanied by another entry, a reference, which renders their object proper also directly accessible, in its capacity as an elemental object, by means of its own physical mark. With the entries belonging to class *c*, on the other hand, though they are characterized by the same duplicity of their principal object, this does not entail any outward expression: as both aspects of the principal object of such an entry are rendered directly retrievable by the very same formal mark, there is no need for, nor any possibility of, preparing a second entry.

Let us mention in parentheses that the lack of uniformity and consistency in cataloguing rules and practices, first of all the differences between the corresponding rules of various codes, come to entail here, once again, in spite of the strictest adherence to using the term 'complex object proper of an entry' in the very same sense, considerable variations in the expanse of the actual coverage of this term. For instance, under a code providing for the employment of uniform marks in all main entries, there are bound to be more main entries having also a complex object proper, than under a code prescribing an application of author entries with uniform headings, but at the same time the employment of a physical, a non-uniform mark in the heading of title main entries. The provisions of the latter code are also inconsistent in themselves, which involves further fluctuations in the concrete coverage of the term 'complex object proper'. So, by all their clarity and uniformity in principle, the actual boundaries of the denotation of this term are really liquid and blurred in practice.

7. WHY MUST WE DISTINGUISH BETWEEN AN 'OBJECT PROPER' AND A 'COMPLEX OBJECT PROPER'? The difficulty we are here faced with – our having to set up, for the purposes of the following analysis, two different concepts of the principal entry object – originates in an essential and widespread feature of international cataloguing practice. A great majority of cataloguers neglect to register a big por-

tion of elemental objects in their elemental capacity: they omit to make these objects retrievable under their actual, their physical marks, and are contented with giving them an entry only under the corresponding uniform marks. For instance many cataloguers describe the editions of the works of an author who is or was using several different names or forms of name, only in main and added entries headed by that author's uniform name. Access to these editions by the marks they actually bear, a majority of cataloguers provide only by means of a reference, which, however, does not describe nor even specify them. This means that many elemental objects are not the principal objects of the sole entry which does contain detailed information about them. They can be granted the status of a principal object, more precisely, of an object proper, only by the short cut proposed by us: by a reversal of the ranking order of the complex object proper and the object proper of class *b* entries. On the basis of our original concept of the principal object they would have to be deprived of this status and relegated to the status of objects always incidental. This would be awkward and illogical, since it is surely the most outstanding, and by no means only an incidental, duty of the catalogue to convey appropriate information about *all* the elemental objects within its range. It is also easy to realize that, for the same reason, a terminology deduced directly from our original concept of the principal object would be liable to become confusing: it would force us to call and class just the most prominent object of a majority of entries as their object only second in rank. This is the main reason for our having resorted to the terminological short cut expounded above, which is based on an inversion of the logical ranking order of the two aspects, or as we might also call them, of the two kinds of principal objects.

Obviously, we should have been spared all these complications if the said peculiarity of our cataloguing practice did not exist. If our catalogues contained a main or an added entry headed by their respective physical marks for all the recorded elemental objects without exception, then in making our inventory of these elemental objects we should be free to ignore all entries headed by uniform marks. Such entries could not contain in this case any information about an elemental object that were not conveyed also by one of the entries having a physical mark for heading; and so we should be allowed entirely to disregard the elemental aspect of the objects of entries with uniform headings, i. e., their object proper, and to consider their role as being confined to fulfilling the second and third catalogue functions, to recording composite objects only. Under such conditions the idea of an employment of two diverse concepts of the principal object could not have even arisen, since our fundamental definition – as it would have deprived no elemental object of its due status of a principal object – would have contrived to satisfy every theoretical need.

8. THE FOUR REASONS FOR PREPARING SEVERAL ENTRIES ON ACCOUNT OF ONE BOOK. In the foregoing we have dealt with the difficulty arising from the fact that a sole entry usually provides access to several different objects, and we have defined, from two different points of view, the principal one among these plural objects; thereby obtaining a tool indispensable to carrying out our classification. In possession of this tool, we may now try to cope with the opposite difficulty, originat-

ing in the fact that in the course of cataloguing a *single book* or other physically separate object of cataloguing *several entries* are often prepared.

Some aspects of this topic we have already touched upon in Chapter V: in respect to them we can now confine ourselves to completing and systematizing what we have said before.

Let us, first of all, systematize the remarks, dispersed in Chapter V, on the diverse motives which may induce the cataloguer to prepare more than one entry when cataloguing a single book.

The *first* of these motives is that a book, or in general, a single physical unit, may unite in itself two or even more *primary* elemental objects of cataloguing; e. g., a volume may be not only a materialization of a detached and in itself complete monograph, but at the same time also a number of a series. As we have already stressed above, both these primary elemental objects should be given separate entries of their own. Both these entries are, in essence, as to their function, main entries, although formally one of them may be, and very often is, disguised as an added entry. We shall discuss this point later in the course of our survey of the objects proper of main entries.

The *second* motive is that many formal marks are capable of being fitted into an alphabetical order at two or more different points. One of the reasons for this is that some marks have equivocal forms (e. g., the prefixed or compound surnames); another reason is that the same meaning is expressed in several different forms (e. g., by the original name and a pseudonym of the same person); a third reason, that a particular mark occurs in several different variants (e. g., a personal name spelled in different ways). Most of these alternative and variant forms of formal marks must be made retrievable by the catalogue, for the sake of securing better access to elemental objects, and in view of the second and third catalogue functions. The simplest case of plural entries belonging to this class is occasioned by an anonymous book displaying a physical title different from the uniform title of the work it contains; this case calls for a duplication of entries only. If both the author's name and the title have one physical variant each, the number of necessary entries increases – at least in principle, but in the case of authors with many books also in practice – to three. If there are several physical variants – or even non-physical ones, which is the case if bibliographical usage keeps wavering between different uniform names or titles – the number of entries to be made for this second reason rises even further.

The *third* reason for making more than one entry on account of a single book consists in the latter possessing several formal marks different in meaning but approximately equally efficient in facilitating its retrieval, e. g., a book with two co-authors, an anthology which names its compiler and may be sought under this name as often as under its title, a publication of a corporate body, etc.

The *fourth* reason, finally, originates in the cataloguer's wish to secure an approach under its own formal marks not only to the whole book, but also to one or the other of the components of its contents, or of its notable, characteristic aspects – e. g., to a play forming a part of a dramatic anthology, to an introductory study, to the name of a translator, etc. – and to turn that component or aspect, thus, into a separate autonomous object of cataloguing.

Within every single batch of plural entries prepared for the first and the fourth reason, each entry has its own object proper, and often also its own complex object or objects proper, distinctly different from those of the other entries in the batch. In the batches prepared for the second reason the object proper of a main or added entry is identical with one among the objects proper of the corresponding reference, while their complex object proper, if any, is identical with one of the components of the complex object proper of the lower part of that reference. The entries in the class *c* batches have only their object proper in common; their complex objects proper, however, they may have in common only so far as the second catalogue function is concerned, but even here not necessarily.

The four reasons for the preparation of plural entries co-operate to produce an infinite variety of different batches, of different combinations of entries, corresponding to the no less infinite variety of individual books. It is not at all unusual that for a single book additional entries should be prepared for all of these four reasons, and often even several for each reason. But it is, perhaps, the second reason which has the greatest entry-proliferating effect: it may call for a re-multiplication of all the plural entries made for the other three reasons.

In spite of its simplifying nature, the above classification is, I think, already apt to convey some impression of the richness and intricacy of the pattern produced by the diverse combinations of entries with different objects proper and complex objects proper, prepared in the course of cataloguing large masses of greatly varying books. But we shall be able to apprehend this intricacy even better yet, if we come to surveying these matters from a less general and less abstract viewpoint, and replace the comprehensive concept 'plural entries' by its one degree more concrete constituents: the concepts 'main entry', 'added entry' and 'reference'.[60] We now pass over to dealing with this task. To avoid an over-complication of it, we are going, however, to disregard one of the factors most efficient in enhancing the intricacy of the pattern: the lack of uniformity of cataloguing practices. We shall confine ourselves to taking only the alternatives adopted by the most widespread practices into consideration.

9. THE OBJECT PROPER OF MAIN ENTRIES IS MOSTLY A PRIMARY ELEMENTAL OBJECT. We come now to examine and to classify the main and the added entries according to their objects proper. The fact that objects proper are always elemental objects implies that the viewpoint of the next sections, devoted to that topic, must be the viewpoint of the first catalogue function. When we shall proceed to consider the complex objects proper of main and added entries, i. e., to classify these entries according to the viewpoints of the other two functions, we shall be obliged further to elaborate the results to be arrived at in the next sections, by splitting up the classes here to be established into further sub-classes.

Obviously, we have to begin our replacement of the concept 'plural entries' by examining the objects proper of main entries.

The principal and by far more important class of main entries is constituted by those of them whose object proper is a primary elemental object of cataloguing, i. e., a physically separate unit – in a great majority of cases a book – as a whole; or else an object composed of several such physically separate units (a multi-volumed

[60] As to the 'see also'-references and 'authority cards' offering no new problems in this context cf. p. 162.

book).This is obvious as the heading of most main entries is a formal mark of such a unit, as a rule the most characteristic one among its formal marks. Parenthetically: if such a unit consists of several physically separate parts acquired and processed by the library at different times, all the separate entries prepared for these parts form only a single complete main entry.

Cataloguers are likely in general to agree with the above proposition concerning the object proper of main entries. The proposition is, after all, an inference drawn directly from the definition of the object proper. Besides, the decidedly most important of all the objects cataloguers have to deal with is unquestionably the book itself as a whole; while the principal of all the entries prepared in the course of cataloguing a book is no less unquestionably the main entry. This exalted position of the main entry is based on the facts that, first, of all formal marks of its object proper it has that particular one for its heading which is apt – on the strength of its own merits as well as of its fitting in best with the whole system of rules – to serve the retrieval of that object most effectively, and under which the latter is supposed to be uniformly located in catalogues and bibliographies; and second, that there exists no cataloguing practice which would allow the amount of information in the main entry to be exceeded by the amount conveyed by an entry of another form, whereas the latter is often exceeded, even very considerably, by the former. Any classificatory attempt correlating the most prominent class of elemental objects not with the most effective class of entries, should be regarded beforehand as a downright mistake.

We had better, however, anticipate here a possible objection. Owing to the lack of uniformity and consistency of cataloguing codes, it often happens that the same book is entered under different headings – so, for instance, a book containing a particular collection is entered once under its title, another time under the name of its personal or corporate compiler. About the object proper of the main entry, in the first case no doubt can arise: this object cannot be anything else than the book containing the collection. But what of the added entry made at the same time under the compiler's name? It seems quite plausible to discern the object proper of that added entry in the compiler's 'work', his intellectual product, looked upon as the detached and independent intellectual constituent of the double-faced phenomenon 'book'. But if one accepted this view, would one not be obliged to stick to it also in the case of a main entry made under the same compiler's name by a cataloguer following the opposite practice?

An answer in the affirmative would imply that the proposition concerning the strict correlation between the primary elemental objects of cataloguing and the main entries is untenable. This is, in my opinion, a sufficient reason for rejecting that answer. But its wholly satisfactory repudiation must also include an answer to the question as to the object proper of the added entry made under the compiler's name. This we shall be able to give, however, only at a later stage, in the course of our discussion of co-ordinate added entries (see §11). Meanwhile we must content ourselves with the plain statement that the objection here tentatively raised will duly prove untenable, and unable to impair the validity of the proposition concerning the strict correlation between books and main entries.

This correlation is so firmly established that it persists reciprocally, in both directions. It implies not only that, with but one rather insignificant exception, main entries must have books, primary elemental objects, for their object proper, but also that each primary elemental object must get a separate main entry of its own.

From the latter half of the rule there is an exception, but only a formal, not a real one. The case mentioned as first in our enumeration of the reasons for preparing several entries for one book – the amalgamation of two (or more) primary elemental objects in a single book – is met by many cataloguers by preparing a main entry for that book only in its capacity as a materialization of the monographical work it contains; whereas in its capacity as a member of a series it is allotted only an added entry. Here a primary elemental object of cataloguing seems to be deprived of the main entry that is its due – which in reality it is not! The added entry in question is filed behind a series entry, a leading card for the whole series, and would be defective or even incomprehensible without this. It constitutes in fact merely an addition to, a continuation of, the series leading card which is, of course, a main entry. So the entry in question, although it is given for economical reasons the form of an added entry, is in reality a disguised main entry, and performs exactly the functions of a main entry.

We conclude our discussion of the object proper of the principal class main entries by pointing out a peculiar fact: the strict correlation between main entry and 'book', which we have stated above, does not imply the existence of a similarly firm correlation between the main entry and the 'work' it refers to.

Just like about the mother of a child, there can never be any doubt about the object proper of a particular main entry: that object proper is always the individual book, the primary elemental object, which that main entry describes. This seems to imply that there cannot be any doubt or uncertainty about the work, either, a formal mark of which that main entry must have for its heading: apparently this work can be only the work constituting the content of the book which is the object proper of that main entry. In a great majority of cases this inference is valid indeed. But there are also exceptions: the main entries of a few peculiar groups of books can be headed by a formal mark of not only a single particular work, but by a mark of either of two different works. Cataloguers agree in assigning in these cases a main entry only to one of the two works, and to the other but an added entry; they do not agree, however, upon the choice of the work that, of the two, shall obtain the main entry. Very few codes include specific rules providing for these cases; the choice is usually ceded to the cataloguer, and so the main entries of the book belonging to these categories are divided more or less equally between the two works. Thus, although the object proper of the main entry is with these categories of books, too, quite unequivocally fixed, its complex object proper vacillates between two different alternatives. This will be examined in detail later on (cf. pp. 165–166).

10. A SECOND CLASS OF MAIN ENTRIES IS FORMED BY A SUB-CLASS OF ANALYTICAL ENTRIES. Now we come to examining the second, comparatively insignificant, class of main entries, the existence of which we have already hinted at. This class is constituted by the *analytical entries* in the traditional sense of the word, which resemble in their structure the main entries of those editions of monographs which are published as volumes of series.

As to their object the two sorts of main entries differ essentially: the analytical ones do not record a physical unit in its entirety, only a section of such a unit. So their object proper is not a primary elemental object; which is the reason why they must be regarded as forming a separate class of main entries, essentially different from their principal class. But their object proper is not a secondary elemental object either, since it has no inferior position in relation to a primary elemental object. Nor has the entry itself an accessory character: it is not attached to another entry which records a primary elemental object.

The objects of the analytical entries in the traditional sense occupy thus an intermediate position between the primary and secondary elemental objects. As for the place of these analytical entries themselves within a classification of entries, this cannot be anywhere else than in the class of main entries. 'Main entry' and 'added entry' are categories of entry form ; consequently the question under which of the two analytical entries should be subsumed, must be answered according to the latter's formal structure. Now, this entry conforms with the formal structure of main entries: it is of one cast, and not split up, like the structure of added entries, into two sharply separated parts, with separate objects and 'headings' of their own.

This point we have had to enlarge upon merely because till now cataloguing terminology has entirely neglected to clarify it. As a striking proof of the terminological confusion still prevailing in this respect, I cite the bewilderingly obscure definitions of 'main entry' and 'added entry' in the Glossary to the *AACR*.

We must, however, not omit to mention that by classifying that class of analytical entries we have been here talking about as main entries, we do not wish by any means to exclude the employment of the terms 'analytical entry' and 'analytical cataloguing' with another, broader, denotation. As stated above,[61] we do not consider it inadmissible to invest these terms with a denotation including all accessory, and perhaps even also all tertiary added entries; that they do not refer to a whole work, but only to a particular section or aspect of a work, is no sufficient reason for doing so. The adoption of such an extension of the meaning of these terms would, however, be liable to lead to misunderstandings if not connected with distinguishing between 'analytical main entries' and 'analytical added entries'; the first term being employed with the more widely adopted traditional denotation of the term 'analytical entry', the second as a comprehensive term embracing all accessory and tertiary added entries.

11. A CO-ORDINATE ADDED ENTRY HAS THE SAME OBJECT PROPER AS THE MAIN ENTRY. We come now to the most extensive section of this appendix: the section devoted to added entries. The use of added entries is considerably more varied than that of the other two forms of entry. One might illustrate their comparative roles by saying that at the two ends of the whole field covered by the entries of the author-title catalogue, the main entry and the reference occupy two almost homogeneous sectors, while the midfield, divided into numerous, more or less sharply delimited sub-sectors, is committed to the charge of the added entry.

[61] Cf. p. 77.

Let us begin our survey of the applications of added entries with the statement that all the above mentioned four reasons for preparing several entries on account of a single book (cf. pp. 152–153.) include also the need for preparing added entries. If applied for the first and second reasons, they may even replace main entries and references, respectively; this topic we shall take up later on. The main fields of their use, those in which their services attain the greatest importance, lie, however, in the provinces of the third and fourth reasons. We shall turn our attention now to the added entries made for these two reasons.

Let us repeat that the third reason for making additional entries consists in the advisability of rendering a book retrievable under all its principal formal marks which are approximately equally effective. In § 41 we have closely scrutinized the added entries made for this reason – which we have denominated as'co-ordinate added entries' – and shown that their object proper is identical with the object proper of the main entry, which is the book itself, the primary elemental object of cataloguing as a whole. We had better reinforce this proposition now by refuting an objection which at a first glance might appear plausible.

If for a book a single entry only is prepared, say, a personal author main entry, there can be no doubt about the object proper of that entry: this object cannot be anything else than the primary elemental object (e. g., the book) itself, this being the object which the catalogue has to record first of all and in any case. If, however, for a book of hidden authorship, along with the main entry made, just as in the former case, under the name of the author, also an added entry is prepared under the title, then it is possible to suppose that the latter is liable to have its own object proper different from that of the main entry – a problem another aspect of which we have already met in examining the objects proper of main entries.

But what could this second object be? I cannot think of anything else than the work contained in the book, i. e., the intellectual side, the intellectual constituent of the primary elemental object which is the object proper of the main entry. But it would have to assume this role, then, in the capacity as an independent, self-contained unit, and not in that as a component of the corresponding composite object of the second catalogue function, because in the latter capacity it is not an elemental object, which the object proper of an entry must by definition always be.

To me such a construction would appear artificial and unacceptable. The functions of the catalogue do not offer any justification for recording a particular edition of a work in three different capacities: apart from its capacity as a primary elemental object, also in *two* different capacities as an intellectual object: along with the capacity as a component of one of the composite objects of the second function, also in the capacity as a detached intellectual unit. Such a construction would also be contradicted by generally adopted cataloguing practice. Most codes and cataloguers content themselves with making for a book with an overt personal author a single entry only, and do not even think of making a second entry under the title, i. e., they refrain from making an entry which might eventually be regarded as made for the purpose of recording an elemental object of purely intellectual character, an object different not only from the primary elemental object, but also from the component of the composite objects, concerned. But if they refrain from making an entry for this purpose in the case of a plain author entry, then it is evident that the title added entry they do make in the case of a hidden author main entry, is not made for this purpose either.

So, the only tenable interpretation of the nature of the title added entry emerging in our example, can after all be but the one we have indicated in § 41: it is an entry made solely for the purpose of assisting the main entry in providing adequate access to the primary elemental object which is its object proper, and not for recording a second elemental object. In short, it *is* in fact a co-ordinate added entry.

As to the selection of the heading of the main entry out of the approximately equally effective formal marks of a primary elemental object,international practice is very far from being uniform; often one cannot speak even of a prevalence of a particular one among the diverging practices. It is, for instance, dubious whether it is the cataloguers entering a collection under its compiler who constitute the majority, or those who enter it under its title. The same applies also to most kinds of corporate publications, and to several other bibliographical situations calling for a preparation of co-ordinate added entries. This lack of uniformity is, however, so far as the first catalogue function is concerned, entirely irrelevant to our subject matter: as the objects proper of the main entry and of the added entry co-ordinate with it are in fact identical, the differences in the distribution of the roles and of the formal marks of their identical object proper between these two entries cannot have in this context any relevance.

This being so, one must not wonder at the persistence of the lack of uniformity of cataloguing practice concerning this distribution: in consequence of the approximate equivalence of the two (or more) concurrent formal marks in each of the pertinent cases, there has been (at least so far!) simply no cogent reason for the uniform adoption of a particular one among the respective possible alternatives. The persistent delay in achieving international uniformity in this respect, is in fact a practical proof of the correctness of our view on the nature and function of co-ordinate added entries. Besides, the circumstance that this lack of uniformity is making its appearance with almost every type of added entry which we have classed as co-ordinate, can be assessed as a proof for the soundness of the specification we have given of these types.

Although we have already emphasized it in another context, it is perhaps not superfluous here to repeat that the choice of the main entry heading between two approximately equally effective formal marks is indifferent only from our viewpoint, from the viewpoint of a classifier of entries and objects, but not by any means from that of a rule-maker. One of the principal duties of the latter is to prevent inconsistent, divergent practices. So with regard to the use of the 'equivalent' formal marks in question, he must always make a clear-cut choice; and this choice must be based on a careful consideration and comparison of the respective merits of those 'equivalent' marks. The 'approximative' or 'comparative' equivalence of two different formal marks is a high-grade generalization, admissible on a certain level of a theoretical train of thoughts, but not in practical work, and not in rule-making. With his practical task the codifyer cannot allow himself to depart as far from reality. After a careful examination he must find with regard to each type of case under consideration that one of the alternatives 'approximately equivalent' for the theorist, is for him decidedly preferable to the others.

12. **Accessory and tertiary added entries. Tertiary added entries are separated from co-ordinate ones by a liquid boundary line.** Now we come to the added entries prepared for the fourth reason for making several entries on account of a single book. This reason consists, we repeat, in the wish to raise along with the whole book also sections of it, components of its contents, or else specific aspects of it, to the rank of separate autonomous – *secondary* or *tertiary* – *objects* of cataloguing, by rendering them retrievable under their own formal marks. The reasons for our regarding these two classes of objects and the corresponding two classes of entries – the *accessory* and *tertiary added entries* – respectively, as two separate and distinct classes, have been explained in detail already in Chapter V. But there we have also mentioned that the line of division between co-ordinate and tertiary added entries is liquid to a certain extent, and we promised to examine this fact more closely in an appendix. Our task now is to fulfil that promise.

The criterion by which we have fixed the dividing line between co-ordinate and tertiary added entries is of a quantitative character, and of the inadequate but nevertheless fairly common sort which does not fix the critical quantity with sufficient precision. We shall illustrate this by examples.

Let us take a book containing a collection of folk-tales and revealing, along with the name of the compiler and of the editing folklore society, also that of the translator of the tales. If the book is entered under its title, with three added entries for the said participating agents, no doubt seems to arise about the compiler's added entry having to be classed as a co-ordinate entry, and the translator's as a tertiary one, while opinions on the classing of the entry for the editing body may be divided. The reasons for such a classification of the three entries are the following. Since a title main entry alone does not secure an adequate approach to a book, the cataloguer is bound to prepare at least one additional entry. The luckiest choice he can pick is to make an added entry under the name of a person who had an outstanding role in bringing about the contents of the book; i. e., in our case under the name of the compiler. A somewhat less advantageous possibility is to prepare an entry under the name of the editing learned society; while an entry for the translator is the least fit for this purpose.

Even this example exhibits already a point – the dubious position of the entry under the editing body – which demonstrates manifestly enough that our criterion is liable occasionally to fail owing to its not determining clearly the boundary line which separates co-ordinate from tertiary added entries.

Let us now see what happens if we modify our example by supposing that the book does not communicate either the compiler's name, or that of an editing body. In such a case the position of the translator entry is totally transformed. While originally there was no doubt about its having to be classed as a mere tertiary added entry, now it is improbable that much doubt should arise about its co-ordinate character: many cataloguers will be inclined to consider it a fairly good substitute for the wanting personal compiler entry. (But naturally even in this case it does not lose its separate tertiary object, the translator's performance, either; it merely assumes additionally also the function of a co-ordinate added entry.)

As another example, let us take a book containing the law reports of a high court, entered under the name of the court, with an added title entry, and another added entry for the editing body which is not the court itself. If this body is a ministry of justice, and the collection embraces recent decisions of the court, the latter entry is, rather clearly, a tertiary added entry, seeing that the two entries headed by the name of the court and the title of the book, respectively, suffice to secure adequate access to the latter, so that for this purpose there is no real need for the entry under the ministry. The lower rank of the added entry for the editing body becomes, however, much less certain, if that body is not a ministry, but a law society. And if, besides, the collection contains the law reports of a 17th century court, an added entry under the name of the editing learned society may with every right be considered as equal in efficacy to the main and the title added entry in facilitating the retrieval of the book itself; which means that it is to be ranked as a co-ordinate added entry.

Our examples yield, thus, not only the conclusion that the boundary between co-ordinate and tertiary added entries may, in fact, occasionally become unstable, but also that it is sometimes quite impossible to classify properly the entries belonging to these two classes on the ground of considering only their object proper. It becomes often essential for this purpose to consider also all the concomitant circumstances of the given bibliographical situation.

In concluding this demonstration, we must moreover point out that our schematic examples do not, of course, adequately represent reality: they do not give an idea of the real measure of the liquidity of the border-lines under consideration. Both of them exhibit only a few conspicuous points on a continuously graded scale composed of numbers of slowly, slightly changing situations, and cannot therefore convey an appropriate idea of the numerous occasions for divergent opinions on the right location of these border-lines.

But if it is impossible to draw a clear-cut line of division between them, ought we not to raise the same question with regard to the classification of these two kinds of entries which in Chapter V we have already raised with regard to the classification of their objects? Would it not be more appropriate theoretically, to drop the distinction between co-ordinate and tertiary added entries altogether, and to divide added entries into two fundamental classes only: one constituted by the entries referring to a primary elemental object as a whole, and another composed of the entries recording only a part of such an object?

The answer must be, in my opinion, decidedly in the negative for two different reasons, which also apply to the classification of the objects of these entry classes.

First: the task we are just now occupied with, the classification of entries – as well as the classification of their objects – is primarily a task of theoretical nature. If our end were mainly practical, if we aimed at results to be used as an immediate basis for practical decisions, for instance for formulating cataloguing rules, then we should have to give the opposite answer. For such a purpose concepts, categories and terms with a not entirely precise and clearly delimitable meaning must not directly be used, because they are liable to introduce vagueness, inconsistency and confusion into codes and catalogues. This does not hold with regard to a classification of essentially theoretical character. A classifier is not only permitted but directly obliged to register the two opposite poles formed by the two terminating points of a scale of phenomena, as two distinctly opposite categories, even if the scale exhibits an unbroken, continuous gradation. He must, of course, point out that the exact point where the scale divides into two halves which are to be considered as opposed to each other cannot be fixed with precision, as it is only very gradually that where the two opposites imperceptibly melt into each other. But notwithstanding this, he must not refrain from recognizing the two opposite categories and from locating them in his system. Between the respective functions of a title added entry for a book of hidden authorship and a translator's added entry prepared for a book with a personal author main entry, there is indeed a great difference. A classifier cannot under any conditions be allowed to ignore such a difference and to commingle two things differing so strikingly, even if the dividing line between them is not clearly distinguishable.

Second: Our proposed classification attains a more advanced stage than the contrary one would. A dropping of our class of tertiary added entries would imply also the dropping of our class of tertiary elemental objects. Such a step would be, however, tantamount not only to getting stuck on a one-grade-lower level of classification, but also to distinguishing between elemental objects on the basis of a sole criterion only, at that of an entirely physical nature: the criterion of physical separateness. By dividing the added entries relating to physically separate objects into two distinct classes: those having such an object in its entirety, and those having only one of the aspects of such an object, for their object proper, the classification proposed by us takes a notable step forward. An abandonment of this additional step would rob it of an important and valuable distinction, and subsume entries approximately equivalent to the main entry and entries of an undeniably inferior significance under the same category.

13. THE RELATION OF THE OBJECT PROPER OF ADDED ENTRIES TO THE OBJECT PROPER OF THE MAIN ENTRY. Hitherto we have spoken of added entries as if all of them were linked directly to the pertinent main entry, and their object proper were always linked immediately to the primary elemental object forming the object proper of that main entry. Even our designations of their different classes are based on such an outlook. Now we must take a step forward in our classification of 'added entries', by stating that each of their hitherto mentioned three main classes divides into two sub-classes. In each class it is only the entries belonging to the larger and more important sub-class which are attached directly to the main entry; while those constituting the lesser sub-class are attached immediately to an accessory added entry, and their object proper is attached to a secondary elemental object, and connected with the primary elemental object only indirectly, through the medium of that secondary elemental object.

An example: For an anthology of French drama translated into English added entries are prepared – among others – for the two joint authors and for the translator of one of the plays contained in the anthology; and also for an introductory study to the same play written by an English critic. By rendering our analysis of the relations between these entries more acute, we may disclose the existence of a series of new sub-classes of entries, and make, thus, our classification more discriminative. Such an analysis reveals the added entry under the name of the second-mentioned joint-author as a co-ordinate added entry, but co-ordinate not with the main entry, but only with an accessory added entry, and to be classified therefore as a second-level co-ordinate added entry; the entry for the introductory study as an accessory added entry, linked, again, not directly to the main entry, but to an accessory added entry, and to be classed, thus, as a second-level accessory added entry; and, finally, the translator entry as a tertiary added entry attached to an accessory one, i. e., a second-level tertiary added entry. The objects proper of the two latter entry categories may be called, accordingly, second-level secondary and tertiary elemental objects.

14. THREE INFERIOR CLASSES OF ADDED ENTRIES. Apart from the three principal classes of added entries (subdivided in the above defined way), there are also three specific classes, the existence of which, though they are of minor importance only, we must not ignore.

Two of these classes are applied in overlapping parts of the fields of two of the principal entry classes; the entries in these fields have a hybrid character. The first of these hybrid classes has been dealt with already in the course of our discussion of main entries; it consists of the series added entries made for books pertaining to a series and entered under the formal marks of the self-contained monographic work which forms their content. We have seen that the added entries of this kind are in fact charged with the functions of main entries; they are added entries only formally.

The second of the two hybrid classes of added entries intrudes into the area of references: as regards their real function these added entries are in fact references. This class will be examined later on, together with the reverse class of hybrid entries: with the references actually fulfilling the function of an added entry.

Finally, we should mention that many catalogues contain also added entries of a kind entirely out of place in the author and title catalogue. Not a few codes include rules providing for the preparation of added entries for persons or for works which form the subject-matter of a primary or secondary elemental object. Such added entries are actually subject entries; their headings are simply subject headings. These rules and practices miss the mark, as they are mixing up the functions of the author-title catalogue with those of subject cataloguing. Nevertheless, we must not omit to mention this class of added entries, as they are in fact provided for by many codes confined merely to author-title (and not dictionary!) cataloguing, if only owing to a slip in logic.

15. SOME REMARKS ON REFERENCES. Before concluding the theme of the objects proper of the three principal entry classes, we must say a few supplementary words about the third of these classes: the references. The fact that there are three formal sub-classes of references: one-way references, 'see-also'-references, and 'authority cards', we may here quickly dispose of. Their formal differences are irrelevant to our subject, while as to their objects proper and complex objects proper there are no essential differences between them.

From a classificatory point of view, references resemble main entries in one respect: most of them belong to a single class, which dwarfs their only other class. Their principal class is constituted by the references in the common sense of the word, which have their origin in the fact that a very great portion of the formal marks of books and works occur in several variants. Most of these variants must be recorded by catalogues mainly, but not exclusively, because of the trebling or doubling of the objects of cataloguing entailed by the triplicity of catalogue functions. The reference is a device for rendering objects proper in an economical manner accessible also by means of the formal marks of the corresponding complex objects proper and *vice versa.*

The elemental or composite character of the principal objects of references we have already examined in our classification of entries according to the nature of their objects, and seen that on this basis they may be divided into three classes (cf. pp. 148).

The classification of references according to their objects proper does not give much trouble. The principal object of most references consists in all of their three classes, of a batch of individual objects, which means that their object proper is composed of several elemental objects. There are, however, also a few class *d* and class *e* references which are made on account of a single object only, their object proper being a single elemental object. As for the complex objects proper of references, it may be mentioned that class *d* references have no such objects.

16. ADDED-ENTRY-FUNCTIONED REFERENCES AND REFERENCE-FUNCTIONED ADDED ENTRIES. A small minority of references is constituted by entries which, notwithstanding that they assume the form of a reference, actually play the part of an added entry. We shall now connect the discussion of this kind of references with that of the reverse phenomenon mentioned already in our survey of added entries: the added entry playing the role of a reference. These twin species of entries form two out of the three kinds of hybrid entries, originating in an overlapping of the fields of application of two principal entry classes.

The discussion of the two reciprocal classes of hybrid entries with which we are here concerned presupposes a concise recapitulation of the proper roles, the genuine functions, of added entries as well as of references.

The proper role of added entries may be summarized as consisting, first, of utilizing the formal marks equal in function and retrieval value to, but different in denotation from, the mark employed in the heading of the pertinent main or added entry (co-ordinate added entries); and second, in recording, under their own formal marks first of all on their own account and also for the sake of improving the retrievability of the primary elemental objects as well as the fulfilment of the second and third functions, separate parts or aspects of primary elemental objects (accessory and tertiary added entries). The proper role of references, on the other hand, consists of providing additional access points either to the objects proper or to the complex objects proper of main and added entries, by rendering also the variants and variant forms of the formal marks retrievable which form the headings of these entries.

The border-line between the proper roles of added entries and references, drawn in this way, is transgressed by both the above-mentioned reciprocal hybrid entry classes. The recording of a translator's work is, for instance, undoubtedly a task belonging normally to the range of functions of added entries. Nevertheless, in the case of a translation published in ten editions, the cataloguer may substitute a single reference for the ten added entries. Conversely, many cataloguers, when dealing with the translation of a scholarly work, which probably will not appear but in a single edition, or else with a book displaying on its title-page the name of the author in a wrongly spelled form, do use an added entry for making accessible that (original or translated) form of the title which they relegate in the main entry to a subordinate position, or else the physical but wrong form of the author's name, quite irrespective of the fact that their task consists in both cases of bridging the gulf between a physical and the corresponding uniform mark, a task falling normally upon references.

We wish to emphasize parenthetically that we join the Paris *Statement of Principles* in maintaining that a transgression, in either direction, of the line dividing the 'proper' fields of application of these two forms of entries cannot on any account be disapproved. Added entry and reference are mere technical devices, the range of the use of which should be determined on the grounds only of practical suitability, and not according to principles, to rigid theoretical or classificatory considerations. For the sake of greater clarity and flexibility of our terminology, however, we had better give them specific denominations when used not in their own, but in the opposite field – say, perhaps, the denominations 'added-entry-functioned reference' and 'reference-functioned added entry'.

17. Entries the principal object of which varies with the functions. Hitherto we confined ourselves to classifying entries merely according to their roles in the field of the first function. Consequently, the classes of entries as established above are likely to fit only the objects of the first function without reserve. That they do not prove entirely untenable, and retain at least a part of their validity, when examined from the standpoint of the other two functions, is due to the fact that the complex objects proper of many entries are merely other aspects (the aspects turned towards the second and/or the third function) of their object proper; and that the relations between the members of many batches of entries prepared on account of a single book remain unaltered irrespective of their being surveyed from the side of their object proper or of their complex object proper. In another, still considerable, number of cases, however, our classes turn out to be in need of a qualification, or at least of a subtilization by being split up into sub-classes. As soon as we examine it from the side of the two latter functions, we find that even our terminology as hitherto applied is only limitedly satisfactory. So, for instance, we hit here upon another one-sidedness of, and upon new reservations which have to be made in regard to, our terms 'co-ordinate', 'accessory' and 'tertiary' added entries. Scrutinized from the viewpoints of the last two functions, these three terms are found to be inapplicable to several kinds of entries which, from the standpoint of the first function, can be subsumed under them quite smoothly. Indeed, similar difficulties emerge even with regard to the time-honoured terms 'main entry' and 'added entry'. Although they denote two sharply contrasted categories of entry form, which retain most of their essential differences unimpaired in the spheres of the second and third functions too, the adjectives 'main' and 'added' invest them none the less with connotations which in these spheres no more fit many of them.

Our next task must consist, then, in examining all the changes which our earlier statements regarding, first, the objects proper, and second, the role and significance, of the different entry classes must undergo in view of the fact that the catalogue has not only a first, but also a second and a third function. In performing this task we need, of course, not take into consideration, once more, the simple change of aspects, i. e., the fact that what is the object proper of an entry if it is viewed from the first function, turns into its composite object if it is regarded from the side of the two other functions. This transformation we have already given account of when we have split up the unique principal object of an entry into two of its aspects. Nor have we to heed the entries which are devoid of a complex object proper.

The survey we are going to make will reveal the existence of new classes of entries originating mainly in the fact that many a class splits up into two distinct sections, because the principal object of part of the entries belonging to it remains the same for all the three functions, while the principal object of another part changes with the functions.

We come now to survey one by one the changes, effected by a shifting of our point of view from one function to another, first, in the principal objects of entries, and second, in the significance, the role, of the entry-objects as well as of the entries themselves.

We shall begin our survey with an examination of the relation existing in the field of the second function between a main entry and an added entry hitherto classified as co-ordinate. (A scrutiny of the position of 'co-ordinate added entries' in the field of the third function will have to be carried out separately.)

Seen from the side of the second function, a main entry and the co-ordinate added entries attached to it, have between them, just as when inspected from the side of the first function, a single principal object only. Both kinds of entries are about equally well suited to make accessible not only the primary elemental object they have in common, but also one of this object's reverse sides: the corresponding component of an object of the second function.

This is the general rule, which is, however, subject to – so far as I can see – three exceptions. In § 59 we have already mentioned that there exist three kinds of books (containing a work and a commentary to it, a revised work, or a separately published part of a work) which unite in themselves, or have to be recorded in the capacity of, an edition of two different works, of a component of two different objects of the second function. For being catalogued appropriately, each book of these three kinds has to be allotted one entry for each of the two works concerned. According to generally adopted practice only one of the two entries is a main entry, the other an added entry. From the point of view of the first function, this added entry is to be classified as a accessory, and sometimes even as a tertiary added entry, if its principal object is subordinate in importance to the principal object of the main entry; for instance an added entry made for the person who revised the work to which the main entry is devoted, only slightly, or an added entry made for an original work which has got almost obliterated by a radical revision, to which, accordingly, the main entry is allotted. But if what has been preserved of the original work and what was added to it by the revision are approximately balanced, the main entry may be assigned with equal right to the original and to the revised work. In such a case the added entry which records the other of the two must be classified, from the side of the first function, obviously, as a co-ordinate added entry.

This is the point where the difference makes its appearance when one comes to examining these three groups of cases from the side of the second function. The added entry which from the viewpoint of the first function had to be classified as co-ordinate, cannot be regarded as such any more when examined from the side of the second function. Its principal object is now necessarily and inevitably a component of a composite object of the second function which is distinctly different from that other composite object of this function which the principal object of the main entry forms a component of. This implies that in this special sector of the second function the object of the added entries we are here talking about can never be identical with the object of the main entry; in other words, that here these added entries can never have a co-ordinate character.

We have disclosed here a splitting up of a class of added entries into two distinct sub-classes which we subsumed hitherto indiscriminately under the concept 'co-ordinate added entries', and regarded as a single homogeneous class. In the bigger one of these two sub-classes the principal object of entries *is* a component of the same composite object of the second function as the principal object of the respective main entries, and may rightly be classified as 'co-ordinate' in this respect, too. In the smaller and less significant sub-class – constituted by the three groups of books we are here concerned with – the principal objects of the related main and added entries form components of two different composite objects of the second function, and so from the viewpoint of this the latter cannot be classed as co-ordinate. The denotation of the term 'co-ordinate added entry' is thus, narrower in the sphere of the second function than in that of the first, a part of the first function co-ordinate added entries being excluded from it.

We come now to another change entailed by the shifting of the viewpoint from the first to the other two functions. It differs from the just discussed one in being a change not in the principal object of the entries concerned, but only in the significance, the role of these objects as well as of the entries themselves.

Some of the entries designated above – by virtue of the part they play in the field of the first function – as 'accessory' added entries, lose, so far as the second as well as the third function are concerned, their accessory, i. e., secondary character, in consequence of their principal object being entirely equivalent to the principal objects of those main entries, with which they constitute one composite object of these functions. The most striking example of a clear equivalence of the two kinds of entries as well as of their objects is furnished, on the one hand, by the added entries referring to editions of a particular work which form a part of the contents of a collection, and on the other, by the main entries of the separately published editions of the same work. With regard to the second and third functions, the adjective 'accessory' does not fit the former entries at all. This means, that in the field of these two functions there is no such stable difference of status between main entries and 'accessory' added entries as in that of the first function, where this difference is invariably based on a firmly established relation between the respective objects proper of the two kinds of entries. So we have here another class of added entries which splits into two sub-classes: the class of those entries which we have hitherto called indiscriminately 'accessory'. The entries in one sub-class do retain their accessory character also so far as the second and third functions are concerned, while those in the other do not.

Let us turn now to those instances in which the third function principal object of an entry differs from the same entry's first function principal object.

The most momentous among these differences is that in the area of the third function all entries not headed by the name of a person or body are entirely devoid of all tasks and objects that could lay any claim to being considered and called 'principal'. The object proper of such entries is totally irrelevant to the third function; in other words, so far as this function is concerned, these entries have no object proper, and so, of course, no complex object proper either. The role left to them in this sphere is at best confined to helping a catalogue-user now and again to find his way to an author or body whose name is unknown to him. Accordingly, here they can have only an incidental object (a discussion of which we shall take up later). Even main entries do not make any exception to this: so far as the third function is concerned, main entries with title headings are relegated likewise to the status of entries lacking a principal object.

Here we have come across a factor which splits up every – major as well as minor – class of entries into two sub-classes. The entries in one of these sub-classes – the entries with a personal or corporate name for heading – have also a complex object proper relevant to the third function, whereas those in the other – the title entries – have no such object. So these two kinds of entries form classes opposed to each other not only formally, as regards their headings, but also substantially, in regard to their objects.

An interesting consequence of this is that a divergence in cataloguing rules may here entail more striking classificatory consequences than in any other respect. A rule, for instance, which provides for entering collections under their titles, deprives the main entries for books containing a collection as regards the third function of their principal object, and turns them into entries of only very little importance in this respect; while a rule prescribing that entry should be under the compiler, entrusts the main entries for the same books also with the main task which is to be performed in the sphere of the third function.

A further notable difference is that in the field of the third function added entries completely lose their co-ordinate role. There is in fact no added entry in existence which would be capable of playing the same role in this field which co-ordinate added entries play in the sphere of the first, and mostly also in that of the second function: the role of recording, as their own principal object, the very principal object of the main entry, under a formal mark with a meaning different from that of the formal mark which forms the heading of the main entry. Obviously, reference-functioned added entries are irrelevant to this problem, since the meaning of their heading is not different from that of the heading of the main entry. No less irrelevant are the added entries having not a personal or corporate name for heading, since from the point of view of the third function, as this was just shown, they lack any object that could be considered as principal. We are concerned here only with those not reference-functioned added entries which are headed by a name. These have for heading, however, with only a single exception, always a name different from the name in the heading of the main entry, and so, as regards the third function, their principal object cannot be identical with the principal object of the main entry. The exception mentioned is formed by the author-title added entries, which are, however, even in the field of the first function no co-ordinate added entries.

18. The incidental objects of entries. We come now to our last topic: a survey of the 'incidental objects' of entries and to the classification of entries according to their incidental objects. In the foregoing paragraphs we have been several

times obliged in anticipation to mention this class of entry objects; now we shall examine them more closely.

We propose to use the term 'incidental object of an entry' as a denomination for all autonomous objects which, along with its principal object, that entry may help to follow. To promote the retrievability of its incidental objects is always only a secondary, accessory service of an entry. The incidentality of a connection between object and entry is always reciprocal. If a particular object is only an incidental object of an entry, the entry's significance for that object is also incidental: it plays only a subsidiary role in facilitating the retrieval of that object. The catalogue never fails to contain along with it also another entry – the entry which renders that object autonomous – the principal task of which consists in achieving just the latter end with the highest possible efficiency.

I should like to emphasize that the following survey of the incidental objects of entries is not only an integral part of our classification as a whole, but also an indispensable supplement to, and rectification of, our above representation of the principal objects of entries. This has, as all attempts at classification are liable to have, a certain artificial character. To draw as far as possible clear-cut dividing lines between the classes he thinks fit to establish, belongs to the foremost duties of the classifier. This duty is, however, bound to involve a certain discrepancy between the phenomena of the world of reality. The discrepancy between reality and our attempt at a classification of entries based on the above defined two different concepts of their principal objects is apt to be diminished by the survey of incidental objects of entries which is now following: it will reveal that classes which have been represented in the foregoing as sharply contrasted, are not lacking at all in points of contact.

Among all the entries accessorily and subsidiarily promoting the retrieval of objects which are only incidental to them, it seems natural to mention as first the entries with a referring structure, i. e., the entries split into a top and a bottom section – the references and the added entries – which demonstrate already by their structure the fact that they refer to two different objects.

About references we have already spoken a good deal which must have made it clear enough that if the top section of a reference is a physical mark, then its principal objects are elemental objects, and its incidental object or objects can only be the corresponding one or more composite objects, except class *f* references, or *vice versa*. We are also acquainted with the fact that the alternative in question depends solely on the policy adopted by the given code or library.

As to the added entries, their object proper is, as we know, the object whose formal (physical or uniform) mark they have as their heading. In the case of coordinate added entries this object proper is the book, the primary elemental object of cataloguing, itself, which is, of course, at the same time also the object proper of their bottom section. So, their two sections not referring to two different objects, but only to a single and identical one, in the added entries belonging to this particular class their referring structure does not entail the emergence of any incidental object.

Not so with accessory and tertiary added entries. The bottom section of these refers to the book, but their top section, and so the entry itself, has an object proper different from the book. Consequently, in these entries the book is relegated to the position of an incidental object.

The proposition that the whole book is an incidental object of all accessory and tertiary added entries attached to its main entry, is not a mere fiction, nor an artificial construction, but the statement of a very real and practically important fact. Let us demonstrate this by an example. By the way, this example also intimates the way in which the reality of our proposition on the incidental objects of references which we have just dealt with, may be demonstrated.

The user of the catalogue is looking, say, for an anthology of drama. He does not know either its title, or the name of the compiler, but can remember the author's name of one of the plays the anthology contains. He does not need this particular play at all, but if the catalogue is equipped with an accessory, an analytical entry for that particular play, this entry may provide him with a handle for retrieving the anthology itself, in spite of its object proper being an entirely different one. This shows that the anthology is really one of the objects of that entry, but, of course, only an incidental one.

Doubtless, among the services of accessory and tertiary added entries – even of the second level ones – a prominent place is due to the assistance they offer to the catalogue-users represented in this example. Many cataloguers, when making up their mind in a doubtful case to prepare or to omit an accessory or tertiary added entry, are inclined to think first of all, or even exclusively, about this particular service of these entries; wrongly, of course. They ought to attach at least an equal weight to their two further services: to their making another secondary or tertiary elemental object of cataloguing retrievable by rendering it autonomous, and to their enriching and rounding off those composite objects of which their principal object forms a component.

Between the added entries of these two classes on the one hand, and the main entries on the other, there exists a strict reciprocity in regard to their objects proper and their incidental objects. It is not only that the object proper of a main entry forms an incidental object of all the accessory and tertiary added entries linked to that main entry; but also the objects proper of these secondary and tertiary entries reappear again as incidental objects of the main entry. By the way, with the second of these propositions we have already abandoned the topic of the incidental objects of the entries with a referring structure, and entered the discussion of another class of incidental objects.

That second proposition reveals the existence of a no less real and perhaps even more important kind of incidental objects than the first, of a reverse one. It often happens that a reader in need of a commentary or appendix attached to a book reverts – even though the catalogue does not lack a separate added entry having a formal mark of that commentary or appendix for its heading – to the main entry of the book itself. This happens very probably even more frequently than the previously discussed opposite contingency, that a reader tries to reach a book by using as a handle, instead of the more specific formal marks of the book itself, the name of the author of the commentary or the appendix contained in it. The formal mark in the heading of a main entry is in general more conspicuous, and more likely to attract notice and to be remembered than the formal marks of secondary or tertiary elemental objects forming the headings of accessory or tertiary added entries are. Readers in need of a study on Marlowe which they have once seen in an edition of Marlowe's plays, will more often turn to the heading Marlowe than to the name of the author of the study. Practically, however, this point need not be stressed as strongly as the reverse one, since the preparation of a main entry cannot ever become questionable like the preparation of an accessory or tertiary entry.

A strict reciprocity similar to the one just discussed we come across when comparing the principal and incidental objects of the main and added entries on the one hand, and of the references pertaining to them, on the other. If the former entries have a complex object proper, i. e., an object of composite character as their principal object, this must always form an incidental object of the latter, and *vice versa.* This follows automatically from the function of references. Another aspect of this reciprocity has, by the way, already been dealt with minutely in § 32, in the course of the examination of the secondary services of the two main classes of formal marks.

In concluding our survey of the incidental objects of entries we had better mention, parenthetically, the 'subordinate objects' of cataloguing, too, in spite of the catalogue not rendering them autonomous, i. e., retrievable under their own formal marks, and in spite of their not falling under our definition of incidental objects.[62] Main entries as well as unit card added entries contain a lot of information about this kind of objects, which we mention here for the only reason that our enumeration of the incidental objects should not be left incomplete. We need not, however, enter into a discussion of this topic, as these objects do not involve the preparation of entries of their own, and are, thus, irrelevant not only to the subject of this study, but also to the present appendix.

There is, however, one point regarding subordinate objects which it seems worth while to mention. This is that the entries of many author-title catalogues include information also about a peculiar class of subordinate objects, which are entirely alien to this branch of cataloguing and pertain quite unequivocally to the sphere of subject cataloguing. We are referring to the UDC-numbers and the subject headings – affixed to the entries of these author-title catalogues originally for purposes of processing – under which these entries are filed in the respective catalogues of the library. By means of this 'tracing' the very subjects which the primary elemental object described in the entry deals with, are made subordinate objects of that entry. They may be of substantial help to catalogue-users who have read and enjoyed a book on a particular subject, and wish to read more about it, but do not know where to look for it in the subject catalogues. The fact that in this case the subordinate objects of the author-title-catalogue entries *are* given entries of their own, and made thus autonomous *in other catalogues* of the library, does not, of course, alter their subordinate character so far as the author-title catalogue is concerned.

[62] Cf. p. 22.

APPENDIX III

RECAPITULATION OF THE DEFINITIONS OF THE THREE FUNCTIONS OF STANDARD CATALOGUING

I assemble here the definitions of the three catalogue functions as they have been given and explained above, in accord with the purposes and limits of this study. By doing this I wish to serve the reader's convenience: if somebody may happen to want to look up one of these definitions, he should be provided here with an easier and quicker access to it.

I *a*. The fundamental sector of the first function consists in conveying information about all primary elemental objects, i. e., physically separate items, in the holdings produced by printing or related technics, and using written language as a medium of communication, to which the cataloguer chooses to provide access according to the rules of standard cataloguing.

I *b*. The complementary sector of the first function consists in conveying information optionally

1. about secondary elemental objects, i. e.,

(*i*) about intellectually separate and self-contained components of the complex contents of a primary elemental object (e. g., about the individual papers contained in the report of a symposium), or

(*ii*) about subordinate constituents of the contents of primary elemental objects (e. g., about appendices, accompanying studies, prefaces, epilogues, commentaries, or other explanatory material, illustrations, photographs – the author of all of which is not identical with the author under the name of whom the respective primary elemental object is entered); and finally

2. about tertiary elemental objects, i. e., about specific aspects of primary or secondary elemental objects (e. g., about a translatorial, editorial or reviser's performance of a person different from the author, or about a minor participation of a body in the production of a primary or secondary elemental object which itself is recorded, by means of either a main entry or a coordinate or accessory added entry, under title, under the name of another body with a role of greater consequence, or, in a single specific case, under a personal author).

II. The second function consists in providing assembled information about all those elemental objects recorded in the catalogue, the non-material constituent of which is either the original or a translation of a particular text, collection of texts, or other document-content, or else a revised version, an enlargement, abridgement, adaptation of, or selections from, this original – whether in the latter's language or in a translation arranged in separate composite units.

III. The third function consists in arranging the information about all the elemental objects recorded in the catalogue under persons' or bodies' names into new, composite informational units, each of which communicates pooled information about that portion of the productions and activities of an individual person or body which is included in those elemental objects.

REPERTORY OF THE DEFINITIONS OR EXPLANATIONS OF NEWLY INTRODUCED TERMS AND OF TERMS EMPLOYED DIFFERENTLY FROM PREVAILING USAGE

Bei Fragen zur Produktsicherheit wenden Sie sich bitte an:
If you have any questions regarding product safety,
please contact:

Walter de Gruyter GmbH
Genthiner Straße 13
10785 Berlin
productsafety@degruyterbrill.com